Universitext

Universitext

Universitext is a series of textbooks that presents material from a wide variety of mathematical disciplines at master's level and beyond. The books, often well class-tested by their author, may have an informal, personal, even experimental approach to their subject matter. Some of the most successful and established books in the series have evolved through several editions, always following the evolution of teaching curricula, into very polished texts.

Thus as research topics trickle down into graduate-level teaching, first textbooks written for new, cutting-edge courses may make their way into *Universitext*.

For further volumes:
www.springer.com/series/223

Michael Joswig · Thorsten Theobald

Polyhedral and Algebraic Methods in Computational Geometry

 Springer

Michael Joswig
Fachbereich Mathematik
Technische Universität Darmstadt
Darmstadt, Germany

Thorsten Theobald
Institut für Mathematik, FB 12
Johann Wolfgang Goethe-Universität
Frankfurt am Main, Germany

Originally published in the German language by
Vieweg+Teubner, 65189 Wiesbaden, Germany, as
"Joswig, M.; Theobald, T.; Algorithmische Geometrie"
© Vieweg+Teubner I GWV Fachverlage GmbH, Wiesbaden, 2008

ISSN 0172-5939   ISSN 2191-6675 (electronic)
Universitext
ISBN 978-1-4471-4816-6   ISBN 978-1-4471-4817-3 (eBook)
DOI 10.1007/978-1-4471-4817-3
Springer London Heidelberg New York Dordrecht

Library of Congress Control Number: 2012955474

Mathematics Subject Classification: 51-01, 13P10, 52-01, 68U05

Preface

Geometry is one of the oldest systemized subdisciplines of mathematics. Due to the growing capabilities of computers, algorithmic approaches assume an increasingly significant role within geometry. Against this background, we understand *computational geometry* in a very broad sense as that part of geometry which is (in principle) algorithmically accessible.

The purpose of this book is to provide a functional access to computational aspects of geometry, based on a broad mathematical foundation. Let us point out that the current text is intended to be introductory. Thus restrictions are inevitable, and the choice of topics is naturally biased by the preferences of the authors.

The first part of the book deals with concepts and techniques which refer to polyhedral (i.e., linearly confined) structures. Its mathematical roots lie in discrete and convex geometry. Our treatment includes algorithms for computing convex hulls as well as the construction of Voronoi diagrams and Delone triangulations. The second part is an introduction to some primary concepts in non-linear computational geometry and develops the relevant techniques from computational algebraic geometry. Here, we focus on Gröbner bases and on solving systems of polynomial equations. The third part of the book is devoted to some selected applications in computer graphics, curve reconstruction and robotics.

A prior concern of the book is to establish interconnections between computational-geometric phenomena and other subdisciplines of mathematics (such as algebraic geometry, optimization and numerical mathematics). To achieve this goal we concentrate on some essential ideas and methods. Moreover, the book offers some insights into the possibilities of current computer software (such as `polymake`, `Maple`, or `Singular`) in this context.

Audience and Required Background

The book is directed towards advanced undergraduates and beginning graduates in mathematics and computer science, as well as towards engineering students who are interested in applications of computational geometry (such as in robotics). The book only assumes common concepts from undergraduate courses in linear algebra and calculus. Additional knowledge in discrete mathematics, optimization, algorithms and algebra is useful, however, the material needed from these areas is developed in the text or—in some cases—collected in appendices.

Aim of the Book

It is not intended to cover all the aspects comprehensively. Instead—starting from computational questions in several current topics in geometry—various entry points to more specialized literature and research directions shall be offered.

In contrast to books on computational geometry which originate from computer science, the aspect of abstract data types (which is often important for efficient implementations) is covered only marginally.

History and Acknowledgments

The present book is a revised and updated translation of the German textbook *Algorithmische Geometrie: Polyedrische und algebraische Methoden*, Vieweg, 2008.

The original version resulted from the authors' courses at Technische Universität Berlin, Technische Universität Darmstadt, and Goethe-Universität Frankfurt am Main. The participants of these courses have provided many stimulating discussions and suggestions.

Some of the pictures are courtesy of Sven Herrmann (Fig. 13.3) and Nikolaus Witte (Fig. 1.1).

The translation has been prepared by Theresa Szczepanski and the authors.

The German version benefited from comments and criticism by René Brandenberg, Peter Gritzmann, Martin Henk, Sven Herrmann, Katja Kulas, Alexander Martin, Werner Nickel, Marc Pfetsch, Cordian Riener, Thilo Rörig, Moritz Schmitt, Achill Schürmann, Dieter Schuster, Reinhard Steffens, Natascha Theobald, Tanja Treffinger, Axel Werner, Claudia Wessling, Nikolaus Witte, Ronald Wotzlaw, and Günter M. Ziegler. Further comments by Benjamin Assarf, Roberto Henschel, Katrin Herr, Sadik Iliman, Kai Kellner, Werner Seiler, Christian Trabandt and Timo de Wolff were very helpful when preparing this version.

We are very grateful to everybody for their contributions.

Darmstadt, Germany   Michael Joswig
Frankfurt am Main, Germany   Thorsten Theobald

Contents

Chapter 1
Introduction and Overview

This book studies geometry methodically from an *analytical*, i.e., coordinate-based, viewpoint. In many settings this approach simplifies the computer representation of geometric data. We shall not confine ourselves to linear problems. This is not only appealing from a theoretical viewpoint, it is also practically motivated by advances in computer algebra and the availability of fast computer hardware.

In Chapter 2 we will lay some mathematical foundations. First, we will introduce the language of *projective geometry*, which is very well suited for many geometric applications. Since this is not usually covered in standard introductory courses in mathematics, we briefly discuss the central concepts of projective spaces and projective transformations. We will also introduce the notion of *convexity* in this chapter.

Our analytical approach motivates the structure of this book. It is centered around questions about algorithms which solve systems of equations and their increasingly complex variations with regard to the required mathematical tools.

1.1 Linear Computational Geometry

Most algorithms described in this book are based on *Gaussian elimination*, a core topic in any linear algebra course. In geometric language Gaussian elimination is a procedure which takes a set of affine hyperplanes, $H_1, \ldots, H_k$, in the vector space K^n as input, where K is an arbitrary field. If

$$A = H_1 \cap \cdots \cap H_k \tag{1.1}$$

the output can be an (affine) basis for A, or simply its dimension.

Our foray through computational geometry begins with the real numbers and the transition from equalities to inequalities. Consider for every hyperplane

$$H_i = \left\{ x \in \mathbb{R}^n : \sum_{j=1}^{n} a_{ij} x_j = b_i \right\}$$

M. Joswig, T. Theobald, *Polyhedral and Algebraic Methods in Computational Geometry*,
Universitext, DOI 10.1007/978-1-4471-4817-3_1,

Fig. 1.1 An example of a
bounded polyhedron in $\mathbb{R}^3$.
This particular polyhedron is
a polytope which is dual to a
zonotope. The belt-like strip
in the middle has several very
thin facets

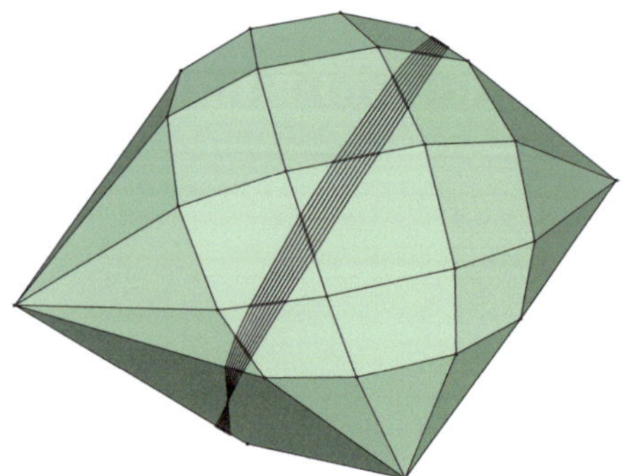

the closed half-space

$$H_i^+ = \left\{ x \in \mathbb{R}^n : \sum_{j=1}^{n} a_{ij} x_j \geq b_i \right\}.$$

The intersection $P = \bigcap_{i=1}^{k} H_i^+$ defines a (*convex*) *polyhedron* (see Fig. 1.1 for an example in $\mathbb{R}^3$).

Polyhedra are fundamental to computational geometry and linear optimization. In higher dimensions, the combinatorial variety of polyhedra is considerably larger than that suggested by lower dimensional images, such as Fig. 1.1. One of the fundamental questions when determining the complexity of many algorithms is, what is the maximum number of vertices that a polyhedron defined by k linear inequalities can have? This question was first answered in 1970 by the *Upper-bound Theorem*. The proof (in a somewhat weaker formulation, see Theorem 3.46) and the explanation of the underlying geometric structure is the first goal of this book. This result is particularly important for computational geometry because we can use it to obtain complexity estimates for several algorithms.

In Chapter 3 we systematically study the properties of polytopes (face lattice, polarity, combinatorics of polytopes) up to *Euler's formula* and the *Dehn–Sommerville equations*. At the end of the chapter we illustrate some of the concepts with the geometric software `polymake`. We will also use this and other software as an aid to understanding the algorithms presented in later chapters.

The core of many mathematical applications is linear optimization, which addresses the problem of computing the minimum or maximum of a linear objective function on a polyhedron P (given by linear inequalities). For computational solutions it is important to note that the polyhedron can be empty, or the objective function can be unbounded on P. In Chapter 4 we give a brief introduction to the relevant aspects of linear optimization. In particular, we discuss the theoretically and practically important *simplex algorithm*. Our main focus (as throughout this text) will be from the geometric perspective.

An interesting computational problem of polytope theory is determining the *entire* set of vertices and rays of a polyhedron defined by a given set of inequalities. Using the duality theory described in Section 3.3, this is equivalent to determining

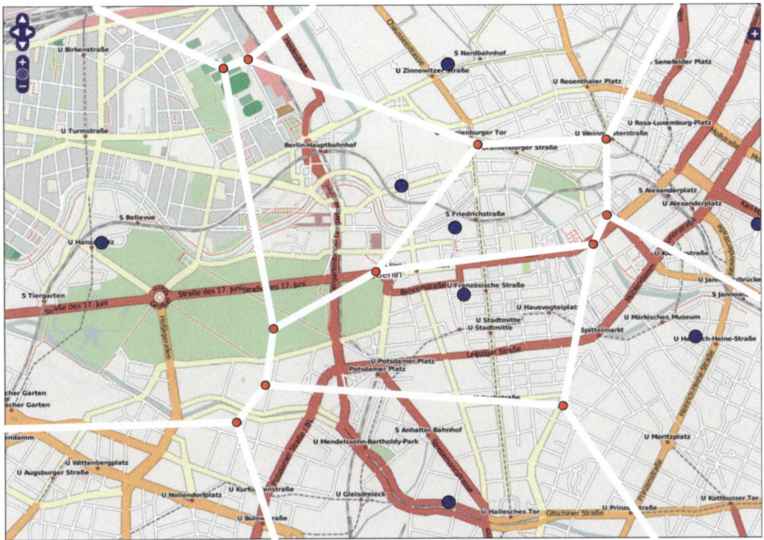

Fig. 1.2 The solution to the post office problem for ten branches of the Deutsche Post AG in Berlin (two of which are not in the visible part of the openstretmap.org map). OpenStreetMap is open data, licensed under the Open Data Commons Open Database License (ODbL). © OpenStreetMap contributors

a minimal system of inequalities which define the convex hull of a point set. We devote Chapter 5 to the *convex hull problem*. For applications it is important to note that actually computing solutions to this problem becomes difficult in higher dimensions (simply because of the large output predicted by the Upper-bound Theorem). A general approach for efficient algorithms is the *divide-and-conquer* principle. We illustrate this by applying it to the computation of convex hulls in the plane.

Next, we examine *Voronoi diagrams* and the corresponding dual *Delone subdivisions*. Given an arbitrary point set $S = \{s^{(1)}, \ldots, s^{(m)}\}$ in the n-dimensional space $\mathbb{R}^n$, the *Voronoi region* corresponding to a point $s^{(i)}$ comprises those points of $\mathbb{R}^n$ which are no further from $s^{(i)}$ (with respect to Euclidean distance) than from any other point of S.

In Chapter 6 we first show how convex hull algorithms can be used to compute Voronoi diagrams in arbitrary dimensions. Afterwards, we concentrate again on the planar case and present the *beach line algorithm*. Knowledge of abstract data types is an advantage for this, so the most important principles will be explained. For a more in depth discussion of common data structures the reader can refer to the recommended literature.

Voronoi diagrams can be used to solve the so-called *post office problem*; a classical application of computational geometry. Given a finite set of points $S \subseteq \mathbb{R}^2$, we efficiently compute for each point $p \in \mathbb{R}^2$ the point $s \in S$ which minimizes the Euclidean distance $\|p - s\|$. The points of S can be interpreted as post offices and the points p as customers. See Fig. 1.2. Of course, one can naively examine every point combination (which is efficient if there is only one customer). However, one

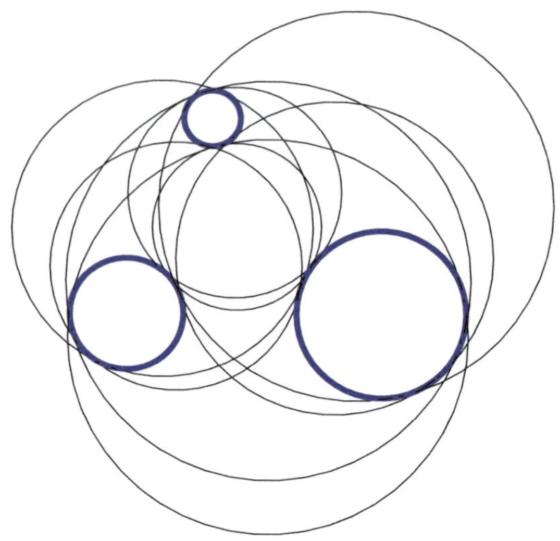

Fig. 1.3 Eight (Apollonius-)circles which touch the three given circles

should interpret the problem as if the postal service wants to create an information system which quickly provides answers for a large group of customers, assuming that the positions of the post offices do not change.

In many applications, the Voronoi diagrams appear in the dual form. Therefore, in Chapter 7 we examine Delone subdivisions and triangulations. A Delone triangulation of the convex hull of a given point set S defined in this manner is in several respects optimal in comparison to all other triangulations of S. We show that in arbitrary dimensions the maximal radius of the circumsphere is minimized. Again, we will examine the planar case in greater detail.

1.2 Non-linear Computational Geometry

The second part of this book is dedicated to non-linear problems. In Chapter 8 we advance from systems of linear equations and inequalities to systems of polynomial equations, and thus into basic algebraic geometry. After this it would be natural to discuss systems of polynomial inequalities, i.e., semi-algebraic geometry, but this would be beyond the scope of this book. At relevant points, we content ourselves with some remarks on polynomial inequalities.

As a good example of a non-linear problem, consider *Apollonius' problem* (Apollonius of Perga ca. 260–190 BC): Given three circles C_1, C_2 and C_3 in the plane, compute another circle that touches each of the previous ones (see Fig. 1.3). If the circles C_1, C_2 and C_3 are in general position there exist eight (possibly complex) solutions. As a possible application, the circles could be interpreted as distance requirements for a set of given points. We will come back to this in later chapters.

In the second part of this book the algorithmic focus is on *Gröbner bases* (Chapter 9). These allow us to solve arbitrary systems of polynomial equations exactly (Chapter 10).

In Chapter 8 we give an introduction to resultants, planar affine and projective algebraic curves and to Bézout's Theorem. We conclude this chapter by illustrating some of these results using `Maple`.

A fundamental algorithmic problem, which is covered in Chapter 9 and will later be the basis of the method we use to solve systems of polynomial equations, is the *Ideal membership problem*. Given polynomials f and $g_1, \ldots, g_r$ in the polynomial ring $K[x_1, \ldots, x_n]$ over the field K, is f in the ideal generated by $g_1, \ldots, g_r$ or not? In general this question cannot be directly answered. This motivates the study of ideal bases with special properties, called Gröbner bases, for which the algorithmic decision problem becomes very simple. Therefore, given a polynomial ideal, the main task is to compute a Gröbner basis for this ideal. We will also develop the relevant theoretical background in computational algebra.

In Chapter 10 we discuss how Gröbner bases are used in the computational solution of systems of polynomial equations. To do this, we first give a brief introduction to the computer algebra system `Singular`. From a theoretical viewpoint, Hilbert's Nullstellensatz plays a fundamental role; it establishes a connection between *geometry* (in the sense of polynomial roots) and *algebra* (in the sense of polynomial ideals). Solutions to systems of polynomial equations may then be obtained from roots of univariate polynomials using elimination ideals. To conclude this chapter we present the most simple case of the *Conti–Traverso algorithm*, which illustrates how to use Gröbner basis techniques in the study of integer linear programs.

1.3 Applications

In the third part of this book we discuss some selected applications of the theoretical results presented earlier.

In Chapter 11 we approach the problem of reconstructing a curve from a given set of points lying on it. We use the concepts of the medial axis and the "local feature size" to evaluate the relationship between the (unknown) curve and the (given) points. The theoretical background from the first part of this book is sufficient for this application.

In Chapter 12 we treat lines in 3- and n-dimensional space. Lines in 3-dimensional space often occur in computational geometry and computer graphics, e.g., in visible surface determination. Although (affine) lines in $\mathbb{R}^3$ are polyhedral objects, questions regarding intersections of lines are intrinsically non-linear. We study these geometric problems by looking at the algebraic characteristics of the *Plücker coordinates* (also known as Grassmann coordinates) of a line. We close this chapter with an example that illustrates the role played by 3-dimensional lines in computer graphics.

Finally, in Chapter 13 we give small insights into applications concerning *Global Positioning Systems (GPS)* and robotics. The functionality of GPS relies on several satellites continuously orbiting the earth so that at least four of them are always accessible from (almost) any position on the Earth's surface. Determining positions using GPS is closely related to a 3-dimensional version of the Apollonius problem, as we will see in Chapter 13. Furthermore, we discuss, sometimes via computer, some fundamental problems of kinematics.

Appendix

In three out of four parts of the appendix we provide foundations for algebraic structures, convex analysis as well as algorithms and complexity. These sections also standardize our notation. The fourth part of the appendix introduces software packages that are used throughout the book: `polymake`, `Maple` and `Singular`. We also mention `CGAL` and `Sage`.

The Structure of This Text   This book consists of more material than a standard one semester course can cover. Hence, this text may be used in several different ways as a basis for a series of lectures. The following compilations are meant as a suggestion:

- "Linear Computational Geometry": Chapters 2 to 7, Chapters 11 and 12. Please note that Chapter 12 uses elimination techniques from Part II of this book. However, the use of `Maple` or `Singular` allows us to treat examples without having a detailed knowledge of the theoretical concepts.
- "Non-linear Computational Geometry": This is complementary to the selection above, hence consisting of Chapters 8 to 10 of the second part of this book and Chapter 13 from the applications part. The amount of material is suitable for a compact course as a follow-up to the course "Linear Computational Geometry".
- "Cross-section of Polyhedral and Algebraic Methods": Chapters 2, 3, 5 or 6, 8 until 10, 12, 13. Sections 9.5 and 10.6 may be left out in this.

Every chapter ends with a small section "Remarks" which references further suggested reading and historical remarks. All figures in this book were produced using the mentioned software and using METAPOST [62].

Part I
Linear Computational Geometry

Chapter 2
Geometric Fundamentals

In this chapter we lay the geometric foundations that will serve as a basis for the topics that we shall meet later. The statements of *projective geometry*, in contrast to those of affine geometry, often allow a particularly simple formulation. The projective equivalence of polytopes and pointed polyhedra (Theorem 3.36) and Bézout's Theorem (Theorem 8.27) on the number of intersections of two algebraic curves in the plane are good examples of this. We will also introduce the notion of *convexity*, which is an irreplaceable concept in linear computational geometry.

2.1 Projective Spaces

The basic motivation behind the introduction of projective spaces comes from the examination of two distinct lines in an arbitrary affine plane, for example the Euclidean plane $\mathbb{R}^2$. The lines either intersect or are parallel to one another. The fundamental idea of projective geometry is to extend the affine plane so that parallel lines have an intersection point at "infinity".

For the remainder of this text, let K be an arbitrary field and for any subset A of a vector space V, let lin A denote the linear hull of A. The cases $K = \mathbb{R}$ and $K = \mathbb{C}$ are of primary interest in this book.

Definition 2.1

(i) Let V be a finite dimensional vector space over K. The *projective space $P(V)$* induced by V is the set of one-dimensional subspaces of V. The dimension of $P(V)$ is defined as $\dim P(V) = \dim V - 1$. The function which maps a vector $v \in V \setminus \{0\}$ to the one-dimensional linear subspace lin v is called the *canonical projection*.

(ii) For any natural number n, the set $P(K^{n+1})$ is called the *n-dimensional projective space over K*. We denote it by $\mathbb{P}^n_K$ and remove the lower index K if the coordinate field is clear from the context.

M. Joswig, T. Theobald, *Polyhedral and Algebraic Methods in Computational Geometry*, Universitext, DOI 10.1007/978-1-4471-4817-3_2,
© Springer-Verlag London 2013

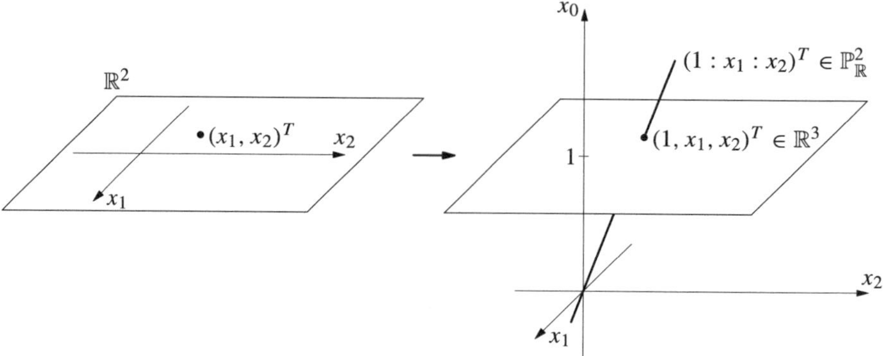

Fig. 2.1 Embedding the Euclidean plane $\mathbb{R}^2$ into the real projective plane $\mathbb{P}^2_{\mathbb{R}}$

A one-dimensional linear subspace U of V is generated by an arbitrary non-zero vector $u \in U$. Thus, we can identify the projective space with the set of equivalence classes of the equivalence relation $\sim$ on $V \setminus \{0\}$, where $x \sim y$ if and only if there exists a $\lambda \in K \setminus \{0\}$ such that $x = \lambda y$.

Definition 2.2 Let $(x_0, \ldots, x_n)^T \in K^{n+1} \setminus \{0\}$ be a vector. Then $x := \lim\{(x_0, \ldots, x_n)^T\} \in \mathbb{P}^n$. We call any element of $x \setminus \{0\}$ *homogeneous coordinates* of x and write $x = (x_0 : \cdots : x_n)^T$, with $(x_0 : \cdots : x_n)^T = (y_0 : \cdots : y_n)^T$ if and only if $(x_0, \ldots, x_n)^T \sim (y_0, \ldots, y_n)^T$, i.e., if there exists a $\lambda \in K \setminus \{0\}$ such that $x_i = \lambda y_i$ for $0 \leq i \leq n$.

We can embed the affine space K^n in the projective space $\mathbb{P}^n_K$ via the injection:

$$\iota : K^n \to \mathbb{P}^n_K, \quad (x_1, \ldots, x_n)^T \mapsto (1 : x_1 : \cdots : x_n)^T. \tag{2.1}$$

Figure 2.1 illustrates the embedding of the Euclidean plane into the real projective plane.

The set of *ideal points* of $\mathbb{P}^n_K$ is

$$\mathbb{P}^n \setminus \iota(K^n) = \left\{ (x_0 : x_1 : \cdots : x_n)^T \in \mathbb{P}^n : x_0 = 0 \right\}.$$

Definition 2.3 Every subspace U of a vector space V defines a *projective subspace* $P(U) = \{\lim(u) : u \in U \setminus \{0\}\}$.

Therefore, the set of (non-empty) projective subspaces of a projective space $P(V)$ is in one-to-one correspondence with the (non-zero) linear subspaces of V. The set of ideal points of $\mathbb{P}^n_K$ forms a subspace of dimension $n - 1$. Also, $\lim \emptyset = \{0\}$ and $P(\{0\}) = \emptyset$.

Projective subspaces of dimension 0, 1 and 2 are called *points*, *lines* and *planes*, as usual. Projective subspaces of dimension $n - 1$ (i.e., codimension 1) are called *hyperplanes*. The embedding $\iota(U)$ of a k-dimensional subspace U of K^n produces a k-dimensional projective subspace called the *projective closure* of U.

Example 2.4 Consider the projective plane $\mathbb{P}_K^2$. The projective lines of this space correspond to the two-dimensional subspaces of K^3. Since the intersection of any two distinct two-dimensional subspaces of K^3 is always one-dimensional, any two distinct lines of the projective plane have a uniquely determined intersection point.

Conversely, given any two distinct projective points there exists one unique projective line incident with both. This follows directly from the fact that the linear hull of two distinct one-dimensional subspaces of a vector space is two-dimensional.

The extension of the affine space K^n to the projective space $\mathbb{P}_K^n$ simplifies many proofs by eliminating case distinctions. In the particularly interesting cases $K = \mathbb{R}$ and $K = \mathbb{C}$, the field K has a locally compact (and connected) topology, inducing the product topology on K^n. This topology has a natural extension to the point sets $\mathbb{P}_\mathbb{R}^n$ and $\mathbb{P}_\mathbb{C}^n$ as a compactification. See Exercise 2.19.

Every hyperplane H in $\mathbb{P}_K^n$ can be expressed as the kernel of a non-trivial *linear form*, that is, a K-linear map

$$\phi : K^{n+1} \to K, \quad x = (x_0 : \cdots : x_n)^T \mapsto u_0 x_0 + \cdots + u_n x_n \qquad (2.2)$$

where the coefficients $u_0, \ldots, u_n \in K$ are not all zero. The set of all K-linear forms on K^{n+1} yields the *dual space* $(K^{n+1})^*$. Pointwise addition and scalar multiplication turns the dual space into a vector space over K. The map ϕ defined in (2.2) is identified with the row vector $u = (u_0, \ldots, u_n)$. Clearly, every hyperplane uniquely defines the vector $u \neq 0$ up to a non-zero scalar and vice versa. In other words: hyperplanes can also be expressed in terms of homogeneous coordinates, and we simply write $H = \ker \phi = [u_0 : \cdots : u_n]$.

The following proposition shows how hyperplanes can be expressed with the help of the *inner product*

$$\langle \cdot, \cdot \rangle : K^{n+1} \times K^{n+1} \to K, \quad \langle x, y \rangle := x_0 y_0 + x_1 y_1 + \cdots + x_n y_n \qquad (2.3)$$

on K^{n+1}. For $x \in K^{n+1}$ and $u \in (K^{n+1})^*$, we write

$$u(x) = u \cdot x = \langle x, u^T \rangle$$

where "$\cdot$" denotes standard matrix multiplication.

Proposition 2.5 *The projective point $x = (x_0 : \cdots : x_n)^T$ lies in the projective hyperplane $u = [u_0 : \cdots : u_n]$ if and only if $\langle x, u^T \rangle = 0$.*

Proof Notice that the condition $\langle x, u^T \rangle = 0$ makes sense in homogeneous coordinates since it is homogeneous itself. The claim follows from the equation

$$\langle (\lambda x_0, \ldots, \lambda x_n)^T, (\mu u_0, \ldots, \mu u_n)^T \rangle = \lambda \mu (x_0 u_0 + \cdots + x_n u_n) = \lambda \mu \langle x, u^T \rangle$$

for every $\lambda, \mu \in K$. $\qquad \square$

At the end of the book, in Theorem 12.24, we will prove a far-reaching generalization of Proposition 2.5.

Example 2.6 As in Example 2.4, consider the affine plane K^2 and its projective closure, the projective plane $\mathbb{P}^2_K$. We can use the homogeneous coordinates to represent a projective line of $\mathbb{P}^2_K$. For $a, b, c \in K$ with $(b, c) \neq (0, 0)$ let

$$\ell = \left\{ \begin{pmatrix} x \\ y \end{pmatrix} \in K^2 : a + bx + cy = 0 \right\}$$

be an arbitrary affine line. Then the projective line $[a : b : c]$ is the projective closure of ℓ. It contains exactly one extra projective point that is not the image of an affine point of the embedding ι. This point is the ideal point of ℓ and has the homogeneous coordinates $(0 : c : -b)$.

The homogeneous coordinates of every line of K^2 parallel to ℓ differ only in a, their first coordinate (in the projective closure). Therefore, they share the same point at infinity. All ideal points lie on the unique projective line $[1 : 0 : 0]$, which is not the projective closure of any affine line. This line is called the *ideal line*.

Ideal points in the real projective plane $\mathbb{P}^2_\mathbb{R}$ are often called *points at infinity* in the literature. The idea of two parallel lines "intersecting at infinity" means that the projective closures of two parallel lines in $\mathbb{R}^2$ intersect at the same ideal point of $\mathbb{P}^2_\mathbb{R}$.

2.2 Projective Transformations

A *linear transformation* is a vector space automorphism, i.e., a bijective linear map from a vector space to itself. Since projective spaces are defined in terms of vector space quotients, linear transformations induce maps between the associated projective spaces.

More precisely, let V be a finite dimensional K-vector space and $f : V \to V$ a K-linear transformation. For $v \in V \setminus \{0\}$ and $\lambda \in K$ we have $f(\lambda v) = \lambda f(v)$ and therefore $f(\mathrm{lin}(v)) = \mathrm{lin}(f(v))$. As f is bijective, non-zero vectors are mapped to non-zero vectors. Hence f induces a *projective transformation*:

$$P(f) : P(V) \to P(V), \quad \mathrm{lin}(v) \mapsto \mathrm{lin}\big(f(v)\big).$$

For $V = K^{n+1}$, the map f is usually described by a matrix $A \in \mathrm{GL}_{n+1} K$. We will therefore use the notation $[A] := P(f)$ for projective transformations. Let $P(V)$ be an n-dimensional projective space. A *flag* of length k is a sequence of projective subspaces $(U_1, \ldots, U_k)$ with $U_1 \subsetneq U_2 \subsetneq \cdots \subsetneq U_k$. The maximal length of a flag is $n + 2$. Every *maximal flag* begins with the empty set and ends with the entire space $P(V)$.

Theorem 2.7 *Let $P(V)$ be a finite dimensional projective space with two maximal flags $(U_0, \ldots, U_{n+1})$ and $(W_0, \ldots, W_{n+1})$. Then there exists a projective transformation $\pi : P(V) \to P(V)$ with $\pi(U_i) = W_i$.*

Proof Since the subspace U_i is strictly larger than U_{i-1}, we can pick vectors $u^{(i)} \in U_i \setminus U_{i-1}$ for $i \in \{1, \ldots, n+1\}$. By construction $u^{(i)}$ is linearly independent of $u^{(1)}, \ldots, u^{(i-1)}$, and therefore $(u^{(1)}, \ldots, u^{(n+1)})$ is a basis of V. Similarly we obtain a second basis $(w^{(1)}, \ldots, w^{(n+1)})$ from the second maximal flag $(W_0, \ldots, W_{n+1})$.

From linear algebra we know that there exists a unique invertible linear map $f : V \to V$ that maps $u^{(i)}$ to $w^{(i)}$ for all $i \in \{1, \ldots, n+1\}$. Therefore $\pi := P(f)$ is a projective transformation with the properties stated in the theorem. $\square$

An equivalent formulation of the above statement is: The group of invertible linear maps $\mathrm{GL}(V)$ operates *transitively* on the maximal flags of $P(V)$.

For a not necessarily maximal flag $\mathcal{F} = (V_1, \ldots, V_k)$ we call the strictly monotone sequence of natural numbers $(\dim_K V_1, \ldots, \dim_K V_k)$ the *type* of $\mathcal{F}$.

Corollary 2.8 *Let* $(U_1, \ldots, U_k)$ *and* $(W_1, \ldots, W_k)$ *be two flags of* $P(V)$ *with the same types. Then there exists a projective transformation* π *on* $P(V)$ *with* $\pi(U_i) = W_i$.

Proof Both $(U_1, \ldots, U_k)$ and $(W_1, \ldots, W_k)$ can be extended to maximal flags. Thus the statement follows from Theorem 2.7. $\square$

One may think that the uniqueness of the linear transformation f in the proof of Theorem 2.7 implies the uniqueness of $\pi = P(f)$. Showing that this is generally not true is the goal of the exercise below. First we clarify some terminology: A point set $M \subseteq \mathbb{P}^n$ is called *collinear* if there exists a projective line that contains all points of M. A quadruple $(a^{(1)}, a^{(2)}, a^{(3)}, a^{(4)})$ of points of $\mathbb{P}^2$ is called a *quadrangle* if no subset of three points is collinear.

Exercise 2.9 For any two quadrangles $(a^{(1)}, a^{(2)}, a^{(3)}, a^{(4)})$ and $(b^{(1)}, b^{(2)}, b^{(3)}, b^{(4)})$ there exists a projective transformation π of $\mathbb{P}^2$ with $\pi(a^{(i)}) = b^{(i)}$ for $1 \le i \le 4$.

An *affine transformation* is a projective transformation that maps ideal points to ideal points.

Exercise 2.10 For every affine transformation π of $\mathbb{P}^n_K$ there exists a linear transformation $A \in \mathrm{GL}_n(K)$ and a vector $v \in K^n$ such that $\pi(\iota(x)) = \iota(Ax + v)$ for all $x \in K^n$.

2.3 Convexity

We begin by summarizing some notation from linear algebra to clarify the terminology and concepts that we will use. As before, let K denote a field.

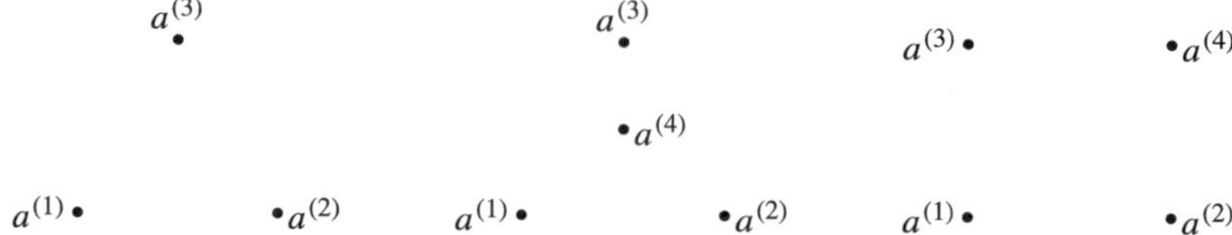

Fig. 2.2 Affinely independent points (*left*) and affinely dependent points (*middle* and *right*) in the Euclidean plane $\mathbb{R}^2$

Definition 2.11 Let $A \subseteq K^n$. An *affine combination* of points in A is a linear combination $\sum_{i=1}^{m} \lambda^{(i)} a^{(i)}$ with $m \geq 1$, $\lambda^{(1)}, \ldots, \lambda^{(m)} \in K$, $a^{(1)}, \ldots, a^{(m)} \in A$ and $\sum_{i=1}^{m} \lambda^{(i)} = 1$. The set of all affine combinations of A is called the *affine hull* of A or simply aff A. We call the points $a^{(1)}, \ldots, a^{(m)} \in K^n$ *affinely independent* if they generate an affine subspace of dimension $m - 1$.

For example, the three points in the picture on the left hand side of Fig. 2.2 are affinely independent and each set of four or more points in the real plane (as in the middle and on the right hand side of Fig. 2.2) are affinely dependent. We set aff $\emptyset = \emptyset$ and dim $\emptyset = -1$.

The language of projective geometry allows us to describe linear algebra over an arbitrary field in geometric terms. In the case of an ordered field like the real numbers (and unlike $\mathbb{C}$) we can further exploit the geometry to obtain results. For the remaining part of this chapter, let K be the field $\mathbb{R}$ of real numbers.

Definition 2.12 Let $A \subseteq \mathbb{R}^n$. A *convex combination* of A is an affine combination $\sum_{i=1}^{m} \lambda^{(i)} a^{(i)}$ which additionally satisfies $\lambda^{(1)}, \ldots, \lambda^{(m)} \geq 0$. The set conv A of all convex combinations of A is called the *convex hull* of A. A set $C \subseteq \mathbb{R}^n$ is called *convex* if it contains all convex combinations that can be obtained from it. The *dimension* of a convex set is the dimension of its affine hull.

The empty set is convex by definition. The simplest non-trivial example of a convex set is the closed interval $[a, b] \subseteq \mathbb{R}$. It is one-dimensional and is the convex hull of its end points. Analogously, for $a, b \in \mathbb{R}^n$ we define:

$$[a, b] := \left\{ \lambda a + (1 - \lambda)b : 0 \leq \lambda \leq 1 \right\} = \text{conv}\{a, b\}.$$

See Fig. 2.3 for some examples.

Exercise 2.13 A set $C \subseteq \mathbb{R}^n$ is convex if and only if for every two points $x, y \in C$, the segment $[x, y]$ is contained in C.

2.3.1 Orientation of Affine Hyperplanes

For real numbers $a_0, a_1, \ldots, a_n$ with $(a_1, \ldots, a_n) \neq 0$ consider the affine hyperplane $H = \{x \in \mathbb{R}^n : a_0 + a_1 x_1 + \cdots + a_n x_n = 0\}$. Then $[a_0 : a_1 : \cdots : a_n]$ are the

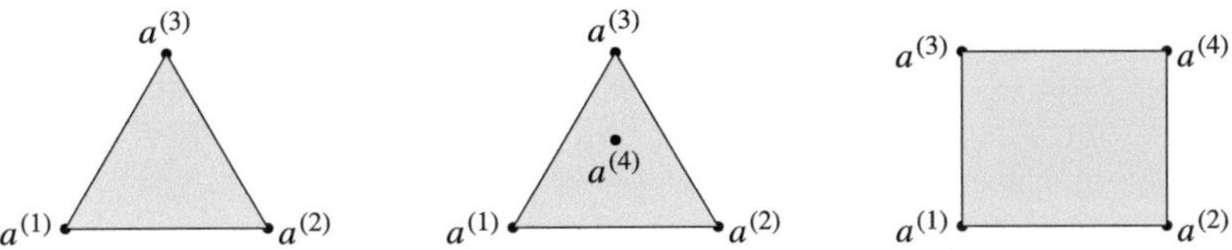

Fig. 2.3 Convex hulls of the points from Fig. 2.2

homogeneous coordinates of its projective closure. The complement $\mathbb{R}^n \setminus H$ has two connected components,

$$H_\circ^+ := \left\{ x \in \mathbb{R}^n : a_0 + a_1 x_1 + \cdots + a_n x_n > 0 \right\} \quad \text{and} \tag{2.4}$$

$$H_\circ^- := \left\{ x \in \mathbb{R}^n : a_0 + a_1 x_1 + \cdots + a_n x_n < 0 \right\}. \tag{2.5}$$

These components are called the *open affine half-spaces* defined by H, with $H_\circ^+$ and $H_\circ^-$ attributed as *positive* and *negative*, respectively. The (*closed*) *positive half-space*

$$H^+ := \left\{ x \in \mathbb{R}^n : a_0 + a_1 x_1 + \cdots + a_n x_n \geq 0 \right\}$$

satisfies $H^+ = H \cup H_\circ^+ = \mathbb{R}^n \setminus H_\circ^-$. The opposite half-space H^- is analogously defined. The vector $(\lambda a_0, \lambda a_1, \ldots, \lambda a_n)$ defines the same affine hyperplane H for any $\lambda \neq 0$, however the roles of H^+ and H^- are reversed when λ is negative. We will let

$$[a_0 : a_1 : \cdots : a_n]^+ := \left\{ x \in \mathbb{R}^n : a_0 + a_1 x_1 + \cdots + a_n x_n \geq 0 \right\}$$

and analogously define $[a_0 : a_1 : \cdots : a_n]^-$. When we wish to distinguish which of the two half-spaces defined by H is positive or negative, we will call $[a_0 : a_1 : \cdots : a_n]$ the *oriented homogeneous coordinates* of H.

We often consider a given affine hyperplane H in $\mathbb{R}^n$ and use the notation H^+ and H^- without having first fixed a coordinate representation of H. This is simply a notational device which enables us to differentiate between the two half-spaces; the coordinates for H can always be chosen so that the notation is in accordance with the above definition.

The inner product introduced in (2.3) is the *Euclidean scalar product* on $\mathbb{R}^n$. As in Proposition 2.5 the sign of the scalar product

$$\left\langle (1, x_1, \ldots, x_n)^T, (a_0, a_1, \ldots, a_n)^T \right\rangle$$

denotes the half-space for $[a_0 : a_1 : \cdots : a_n]$ in which the point $(1, x_1, \ldots, x_n)^T$ lies.

2.3.2 Separation Theorems

For $M \subseteq \mathbb{R}^n$, we let int M denote the *interior* of M. That is, the set of points $p \in M$ for which there exists an ϵ-ball centered at p, completely contained in M. A set is called *open* when $\mathrm{int}\,M = M$ and is *closed* if it is the complement of an open set. The *closure* $\overline{M}$ of M is the smallest closed set in $\mathbb{R}^n$ containing M. The set $\partial M := \overline{M} \setminus \mathrm{int}\,M$ is the *boundary* of M. All of these terms are defined with respect to the ambient space $\mathbb{R}^n$.

Some concepts from analysis are essential for the structure theory of convex sets. The following statements rely on two core results which are proved in Appendix B. Here, an affine hyperplane H is called a *supporting hyperplane* for a convex set $C \subseteq \mathbb{R}^n$ if $H \cap C \neq \emptyset$ and C is entirely contained in one of the closed affine half-spaces determined by H.

Theorem 2.14 *Let C be a closed and convex subset of $\mathbb{R}^n$ and $p \in \mathbb{R}^n \setminus C$ an exterior point. Then there exists an affine hyperplane H with $C \subseteq H^+$ and $p \in H^-$, that meets neither C nor p.*

The next statement is a direct consequence of Theorem 2.14.

Corollary 2.15 *Let C be a closed and convex subset of $\mathbb{R}^n$. Then every point of the boundary ∂C is contained in a supporting hyperplane.*

A convex set $C \subseteq \mathbb{R}^n$ is called *full-dimensional* if $\dim C = n$. When C is not full dimensional, it is often useful to use these topological concepts with respect to the affine hull. The *relative interior* $\mathrm{relint}\,C$ of a convex set C consists of the interior points of C interpreted as a subset of $\mathrm{aff}\,C$. Analogously, the *relative boundary* of C is the boundary of C as a subset of $\mathrm{aff}\,C$.

2.4 Exercises

Exercise 2.16 Let $P(V)$ be a projective space. For every set $S \subseteq V$ the set

$$T = \big\{ \mathrm{lin}\{x\} : x \in S \setminus \{0\} \big\}$$

is a subset of $P(V)$ and for the subspace $\mathrm{lin}\,S$ generated by S, $P(\mathrm{lin}\,S)$ is a projective subspace which we denote by $\langle T \rangle$. Prove the dimension formula

$$\dim U + \dim W = \dim\big(\langle U \cup W \rangle\big) + \dim(U \cap W)$$

for two arbitrary projective subspaces U and W of $P(V)$.

Exercise 2.17 Let K be any field, and let $A = (a_{ij}) \in \mathrm{GL}_{n+1}\,K$. Show:

(a) If H is a projective hyperplane with homogeneous coordinates $(h_0 : h_1 : \cdots : h_n)$ then the image $[A]H$ under the projective transformation $[A]$ is the kernel of the linear form with coefficients $(h_0, h_1, \ldots, h_n)A^{-1}$.

(b) The projective transformation $[A]$ acting on $\mathbb{P}^n_K$ is affine if and only if $a_{12} = a_{13} = \cdots = a_{1,n+1} = 0$.

Exercise 2.18

(a) Every projective transformation on the real projective line $\mathbb{P}^1_{\mathbb{R}}$ (apart from the identity) has at most two fixed points.
(b) Every projective transformation on the complex projective line $\mathbb{P}^1_{\mathbb{C}}$ (apart from the identity) has at least one and at most two fixed points. (Explain why it is natural to talk about a double fixed point in the first case.)

A projective space over a topological field has a natural topology that will be discussed in the following exercise.

Exercise 2.19 Let $\mathbb{K} \in \{\mathbb{R}, \mathbb{C}\}$. Show:

(a) The point set of a projective space $\mathbb{P}^n_{\mathbb{K}} = \mathbb{K}^{n+1}/\!\sim$ is compact with respect to the quotient topology.
(b) Every projective subspace of $\mathbb{P}^n_{\mathbb{K}}$, interpreted as a subset of the points of $\mathbb{P}^n_{\mathbb{K}}$, is compact.

Exercise 2.20 Let K be a finite field with q elements.

(a) Show that the projective plane $\mathbb{P}^2_K$ has exactly $N := q^2 + q + 1$ points and equally many lines.
(b) Denote by $p^{(1)}, \ldots, p^{(N)}$ the points and by $\ell_1, \ldots, \ell_N$ the lines of $\mathbb{P}^2_K$. Furthermore, let $A \in \mathbb{R}^{N \times N}$ be the *incidence matrix* defined by

$$a_{ij} = \begin{cases} 1 & \text{if } p^{(i)} \text{ lies on } \ell_j, \\ 0 & \text{otherwise.} \end{cases}$$

Compute the absolute value of the determinant of A. [*Hint*: Study the matrix $A \cdot A^T$.]

Exercise 2.21 (Carathéodory's Theorem) If $A \subseteq \mathbb{R}^n$ and $x \in \operatorname{conv} A$, then x can be written as a convex combination of at most $n + 1$ points in A. [*Hint*: Since $m \geq n + 2$ points are affinely dependent, every convex combination of m points in A can be written as a convex combination of $m - 1$ points.]

2.5 Remarks

For further material on projective geometry, refer to the books of Beutelspacher and Rosenbaum [13] and Richter-Gebert [88]. More detailed descriptions of convexity can be found in Grünbaum [56, §2], Webster [98] or Gruber [55]. For basic topological concepts, see the books of Crossley [30] and Hatcher [58]. Although our projective transformations are by definition always linearly induced, in other texts it is common to extend this notion to include collineations induced by field automorphisms.

Chapter 3
Polytopes and Polyhedra

Polytopes may be defined as the convex hull of finitely many points in n-dimensional space $\mathbb{R}^n$. They are fundamental objects in computational geometry. When studying polytopes, it soon becomes apparent that the proof of seemingly obvious properties often requires further clarification of the basic underlying geometric structures. An example of this is the major result that polytopes can also be represented as the intersection of finitely many affine half-spaces.

In this chapter the geometric foundations of polytopes and unbounded polyhedra will be presented from a computational viewpoint.

3.1 Definitions and Fundamental Properties

Definition 3.1 A set $P \subseteq \mathbb{R}^n$ is a *polytope* if it can be expressed as the convex hull of finitely many points. A k-dimensional polytope is called a *k-polytope*. The convex hull of $k + 1$ affinely independent points is a *k-simplex*.

A 0-dimensional polytope is just a point, a 1-polytope is a line segment and the 2-dimensional polytopes are precisely the convex polygons. We adopt the convention that the empty set is a polytope of dimension -1. See Fig. 3.1 for some examples.

From an analytical viewpoint, a polytope is a closed and bounded, and hence compact, subset of $\mathbb{R}^n$. Polytopes in lower dimensions illustrate neither the diversity of polytopes nor the depth of higher dimensional polytope theory.

We now introduce the reader to some examples of polytopes which will be useful in the following sections.

The *standard cube* C_n is the convex hull of the 2^n points which have ± 1-coordinates. If we denote the standard basis vectors in $\mathbb{R}^n$ by $e^{(1)}, \dots, e^{(n)}$, then we can express the *cross-polytope* as the convex hull of the $2n$ points $\pm e^{(1)}, \dots, \pm e^{(n)}$. The 3-dimensional cross-polytope is the *regular octahedron*.

The *cyclic polytopes* form an important class of polytopes with extremal properties. These properties will be discussed in further detail in Section 3.5.

M. Joswig, T. Theobald, *Polyhedral and Algebraic Methods in Computational Geometry*, Universitext, DOI 10.1007/978-1-4471-4817-3_3,
© Springer-Verlag London 2013

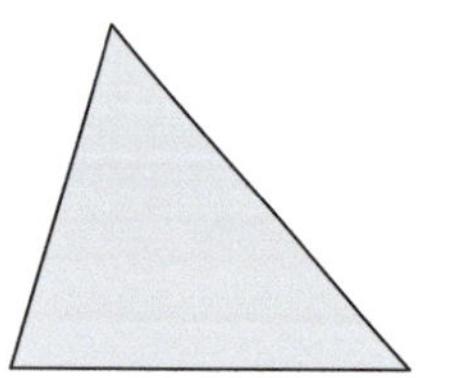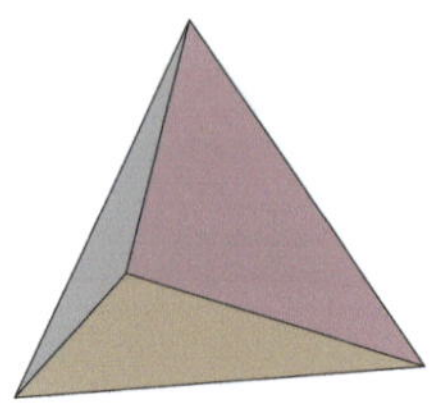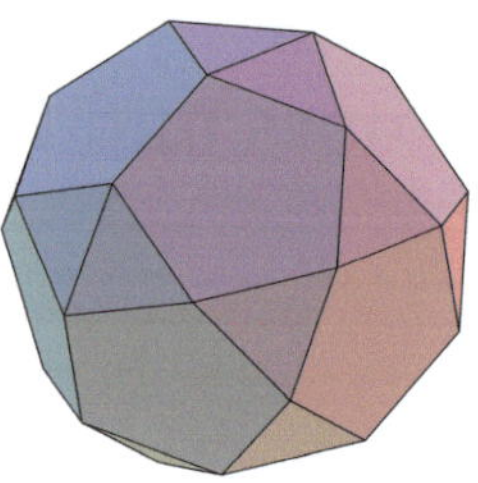

Fig. 3.1 Every 2-simplex is a triangle and every 3-simplex in $\mathbb{R}^3$ is a (generally irregular) tetrahedron. The *right hand picture* shows a 3-polytope in $\mathbb{R}^3$

Definition 3.2 The *moment curve* μ_n in $\mathbb{R}^n$ is defined as

$$\mu_n : \mathbb{R} \to \mathbb{R}^n, \quad \tau \mapsto \left(\tau, \tau^2, \ldots, \tau^n\right)^T.$$

A polytope $Z \subseteq \mathbb{R}^n$ is called *cyclic* if Z is the convex hull of points of the moment curve.

For $n = 2$ the moment curve is the standard parabola $\tau \mapsto (\tau, \tau^2)^T$. Any cyclic 2-polytope that is defined as the convex hull of $m \geq 3$ points is a convex m-gon. This is independent of the specific choice of the m points on the curve μ_2.

It is easy to verify that the image of a polytope under an affine transformation is again a polytope (of the same dimension).

Definition 3.3 An *affine automorphism* of a polytope $P \subseteq \mathbb{R}^n$ is an affine transformation of $\mathbb{R}^n$ that maps P onto itself.

The set of all affine automorphisms of a polytope is a group with respect to composition. The size of this *automorphism group* is a measure of the regularity of the polytope.

Exercise 3.4

(a) Show that for any two points $p, q \in \mathbb{R}^n$ with ± 1-coordinates there exists an affine transformation of $\mathbb{R}^n$ which leaves the standard cube fixed and maps p to q.
(b) Compute the number of affine automorphisms of the standard cube. [*Hint*: Make use of Theorem 2.7.]

3.1.1 The Faces of a Polytope

Based on the concept of supporting hyperplanes (as introduced in Section 2.3.2 or in Appendix B) we define the faces of a polytope.

Fig. 3.2 A cube in $\mathbb{R}^3$ has 8 vertices, 12 edges (which are also ridges in dimension 3) and 6 facets

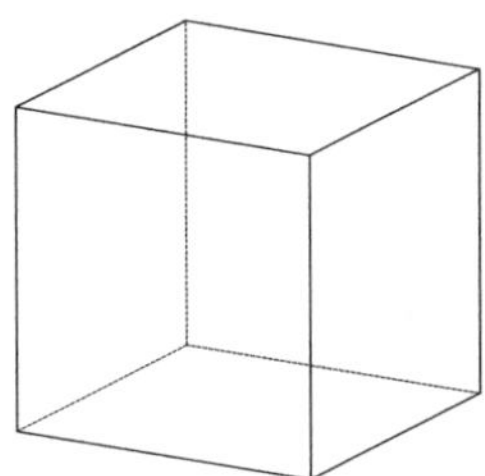

Definition 3.5 Let $P \subseteq \mathbb{R}^n$ be an n-polytope. The intersection $P \cap H$ of P with a supporting hyperplane H is called a *proper face* of P. A face of dimension k is called a *k-face*. A 0-face is called a *vertex*, a 1-face an *edge*, an $(n-2)$-face a *ridge* and an $(n-1)$-face is called a *facet*. Additionally, there are two *non-proper faces*: the empty set and P itself.

Definition 3.5 was stated for full-dimensional polytopes. The terms translate immediately to arbitrary k-polytopes $P \subseteq \mathbb{R}^n$ for $k < n$ if they are interpreted with respect to the affine hull aff P of P. An example illustrating the faces of a polytope is given in Fig. 3.2.

Theorem 3.6 *The number of faces of a polytope is finite. Faces of polytopes are polytopes themselves.*

Proof Let $P = \operatorname{conv} U$ for a finite set U. For both claims it suffices to show that every proper face of P is the convex hull of a subset of U. Let H be a supporting hyperplane of P and let $U' := U \cap H$. The oriented homogeneous coordinates of H are $[a_0 : \cdots : a_n]$ and we assume without loss of generality that $P \subseteq H^+$. We will show that $H \cap P = \operatorname{conv} U'$. The inclusion "$\supseteq$" is clear.

For the reverse inclusion consider a point $p = (p_1, \ldots, p_n)^T \in P \setminus \operatorname{conv} U'$. There exist $u^{(1)}, \ldots, u^{(k)} \in U$ such that $p = \lambda^{(1)} u^{(1)} + \cdots + \lambda^{(k)} u^{(k)}$ with $\lambda^{(j)} \geq 0$ and $\sum \lambda^{(j)} = 1$. Here we can assume that $u^{(1)} \in U \setminus U'$ and $\lambda^{(1)} > 0$. We have to show that $p \notin H$. We have

$$a_0 + \sum_{i=1}^{n} a_i p_i = a_0 + \sum_{i=1}^{n} a_i \sum_{j=1}^{k} \lambda^{(j)} u_i^{(j)}$$

$$= a_0 + \sum_{j=1}^{k} \lambda^{(j)} \sum_{i=1}^{n} a_i u_i^{(j)} = \sum_{j=1}^{k} \lambda^{(j)} \left(a_0 + \sum_{i=1}^{n} a_i u_i^{(j)} \right),$$

where the last equation follows from $\sum_{j=1}^{k} \lambda^{(j)} = 1$. But by our assumption we have that $a_0 + \sum_{i=1}^{n} a_i u_i^{(1)} > 0$ and $a_0 + \sum_{i=1}^{n} a_i u_i^{(j)} \geq 0$ for all $j \in \{2, \ldots, n\}$. Since $\lambda^{(1)} > 0$, this implies $a_0 + \sum_{i=1}^{n} a_i p_i > 0$ or, in other words, $p \in H_\circ^+$. $\square$

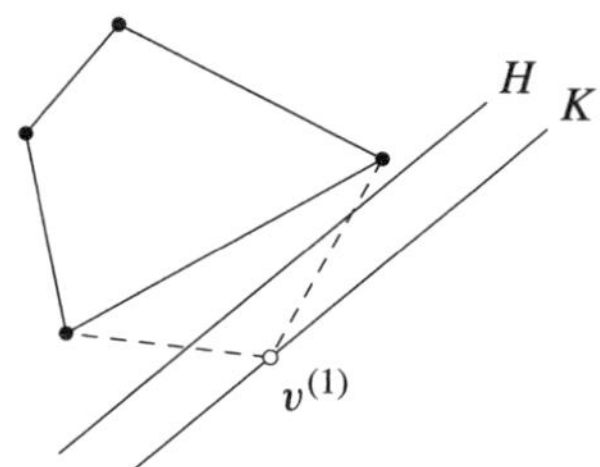

Fig. 3.3 Separation of the point $v^{(1)}$ from $\mathrm{conv}(V \setminus \{v^{(1)}\})$ by H and a parallel supporting hyperplane K

3.1.2 First Consequences of the Separating Hyperplane Theorem

As the Separating Hyperplane Theorem 2.14 is an important key to the study of the structure of polytopes, we shall begin by studying its implications.

Theorem 3.7 *The boundary of a full-dimensional polytope $P \subseteq \mathbb{R}^n$ is the union of all of its proper faces.*

Proof Clearly the union of all proper faces of P is contained in the boundary of P. The reverse inclusion is implied by Corollary 2.15, which states that every boundary point intersects at least one supporting hyperplane. $\square$

Theorem 3.8 *Every polytope is the convex hull of its vertices.*

Proof Let $P = \mathrm{conv}\, U$ for a finite set U. After successively removing all points of U that can be expressed as a convex combination of other points in U, we obtain a subset $V = \{v^{(1)}, \ldots, v^{(k)}\}$ that satisfies $P = \mathrm{conv}\, V$ and which is minimal with respect to containment.

We now show that every remaining point is a vertex of P. It suffices to show this for $v^{(1)}$. Since V was chosen to be minimal, $v^{(1)}$ is not contained in the convex hull of the other points. By Theorem 2.14 there exists an affine hyperplane H that separates $v^{(1)}$ and $\mathrm{conv}\{v^{(2)}, \ldots, v^{(k)}\}$. We set $H = [a_0 : \cdots : a_n]$ and assume that $v^{(1)} \in H_{\mathrm{o}}^-$. Using the notation $\mu := a_0 + \sum_{i=1}^{n} a_i v_i^{(1)}$, the hyperplane K which is parallel to H and contains $v^{(1)}$ has the oriented homogeneous coordinates $[a_0 - \mu : a_1 : \cdots : a_n]$; see Fig. 3.3. The inequality $\mu < 0$ implies $\{v^{(2)}, \ldots, v^{(k)}\} \subseteq \mathrm{int}\, K^+$ and since $v^{(1)} \in K$ we have that K is a supporting hyperplane to P. Now let $p \in P \cap K$. Since p is a convex combination of the points $v^{(j)}$, i.e., $p = \sum_{j=1}^{k} \lambda^{(j)} v^{(j)}$ for appropriate $\lambda^{(j)} \geq 0$ with $\sum_{j=1}^{k} \lambda^{(j)} = 1$, we have

$$a_0 - \mu + \sum_{i=1}^{n} a_i p_i = a_0 - \mu + \sum_{i=1}^{n} a_i \sum_{j=1}^{k} \lambda^{(j)} v_i^{(j)}$$

$$= \sum_{j=1}^{k} \lambda^{(j)} \left(a_0 - \mu + \sum_{i=1}^{n} a_i v_i^{(j)} \right) = 0.$$

Fig. 3.4 The point q is not contained in the affine hull of p with any face of dimension $\leq n - 2$ (here $n = 2$)

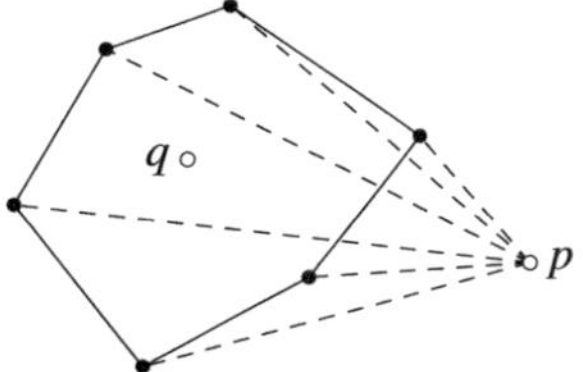

Since $\lambda^{(j)} \geq 0$ and $a_0 - \mu + \sum_{i=1}^{n} a_i v_i^{(j)} > 0$ for all $j \geq 2$, we have $\lambda^{(2)} = \cdots = \lambda^{(k)} = 0$ and also $\lambda^{(1)} = 1$. This means that $p = v^{(1)}$ and therefore that $v^{(1)}$ is a vertex of P. $\square$

An immediate consequence of the above theorem is that the containment-minimal set V of points that generate P is uniquely determined.

3.1.3 The Outer Description of a Polytope

The representation of a polytope as the convex hull of a finite point set is called the *$\mathcal{V}$-representation* or *inner description*. The following two central theorems state that every polytope can be equivalently described as the bounded intersection of finitely many closed half-spaces (the *$\mathcal{H}$-representation* or *outer description*). The prefixes $\mathcal{V}$- and $\mathcal{H}$- derive from the terms "vertices" and "hyperplanes".

Theorem 3.9 *Let* $P \subseteq \mathbb{R}^n$ *be an n-polytope,* $\{F_1, \ldots, F_m\}$ *the set of its facets,* H_i *the supporting hyperplane to P at F_i and H_i^+ the half-space containing P. Then we have*

$$P = \bigcap_{i=1}^{m} H_i^+.$$

Thus every polytope is the intersection of a finite set of closed half-spaces.

Proof The inclusion "$\subseteq$" is clear. For the inclusion "$\supseteq$" we show that every point outside of P is not contained in the intersection $\bigcap_{i=1}^{m} H_i^+$. For the following we fix a point $p \notin P$.

We study the set $\{G_1, \ldots, G_k\}$ of all faces of P of dimension $\leq n - 2$. Let q be a point in the interior of P which is not contained in the set $\bigcup_{i=1}^{k} \mathrm{aff}(G_i \cup \{p\})$. Such a point exists since the interior of an n-polytope has dimension n and can therefore not be covered by a finite number of affine subspaces of dimension $\leq n - 1$ (see Fig. 3.4). The segment $[p, q]$ intersects the boundary of P in a uniquely determined point z which, by Theorem 3.7, is contained in a proper face of P. By the choice of q it is guaranteed that z is not contained in a face of dimension $j < n - 1$. This implies that there exists an $i \in \{1, \ldots, m\}$ with $z \in F_i$. So we have $z \in H_i$ and $q \in H_i^+$, but $p \in H_i^- \setminus H_i$, i.e., $p \notin \bigcap_{i=1}^{m} H_i^+$. $\square$

When, as in Theorem 3.9, the polytope $P \subseteq \mathbb{R}^n$ is full-dimensional, the affine span of every facet F defines a hyperplane H. Assuming that H has the form $H = [a_0 : \cdots : a_n]$ and $P \subseteq H^+$, every positive multiple of $(a_1, \ldots, a_n)^T$ is called an *inner normal vector* of F and every negative multiple of $(a_1, \ldots, a_n)^T$ is called an *outer normal vector* of F. If $\dim P < n$, then for any facet F, there exist infinitely many affine hyperplanes of $\mathbb{R}^n$ that contain F.

We can now refine Theorem 3.7, which states that the boundary of a polytope is the union of its facets.

Theorem 3.10 *If the intersection P of a finite number of closed affine half-spaces in $\mathbb{R}^n$ is bounded, then P is a polytope.*

Proof The proof is completed by induction over the dimension n of the space. The statement is clear for dimension ≤ 1 . So let $n \geq 2$ and

$$P = \bigcap_{i=1}^{m} H_i^+$$

be the bounded intersection of a finite number of affine half-spaces in $\mathbb{R}^n$. Let $F_j := H_j \cap P$, $j \in \{1, \ldots, m\}$. Then F_j is a bounded intersection of half-spaces in the hyperplane H_j. Since H_j can be identified with an affine space of dimension $n - 1$, we know by the inductive hypothesis that F_j is a polytope in H_j and therefore also a polytope in $\mathbb{R}^n$. Let V_j be the set of vertices of F_j and $V = \bigcup_{j=1}^{m} V_j$.

It suffices to show that $P = \operatorname{conv} V$. The inclusion "$\supseteq$" is clear since $V \subseteq P$ and P is convex. For the reverse inclusion consider a point $q \in P$. If q is a boundary point of P, then there exists a $j \in \{1, \ldots, m\}$ with $q \in F_j$. The point q is therefore a convex combination of V_j which in particular implies that $q \in \operatorname{conv} V$. If q is contained in the interior of P, then q is contained in a segment, $[r, s]$, formed by the intersection of a line with P. Since r and s are on the boundary of P they are contained in $\operatorname{conv} V$ and thus $q \in \operatorname{conv} V$. $\square$

Example 3.11 The hyperplanes that define facets of the standard cube C_n are precisely $H_i = [1 : h_1^{(i)} : \cdots : h_n^{(i)}]$ for $i \in \{1, \ldots, 2n\}$ with

$$h_k^{(i)} = \begin{cases} 1 & \text{if } i = k, \\ -1 & \text{if } i = k + n, \\ 0 & \text{otherwise} \end{cases}$$

for $k \in \{1, \ldots, n\}$.

Computing the $\mathcal{H}$-representation when given the $\mathcal{V}$-representation of a polytope and vice versa is a major topic of computational geometry and will be discussed in Chapter 5.

Exercise 3.12 Show that the intersection of a polytope with an arbitrary affine subspace is a polytope.

Exercise 3.13 For a polytope P show that:

(a) The intersection of a set of faces of P is a face of P.
(b) Every ridge of P is the intersection of exactly two facets of P.
(c) If G is a face of P and F is a face of G, then F is a face of P.

3.2 The Face Lattice of a Polytope

Containment defines a partial order on the set $\mathcal{F}(P)$ of all faces of a polytope P. Theorem 3.6 tells us that this set is finite, i.e., $(\mathcal{F}(P), \subseteq)$ is a finite *partially ordered set* (or *poset*). As a purely combinatorial object this poset is an important interface between the analytically focused general complexity theory and discrete geometry.

Exercise 3.14 Show that $(\mathcal{F}(P), \subseteq)$ satisfies the following conditions:

(a) There exists a uniquely determined smallest and largest face of P.
(b) For two arbitrary faces $F, G \in \mathcal{F}(P)$ there exists a uniquely determined smallest face $F \vee G$ such that $F \subseteq F \vee G$ and $G \subseteq F \vee G$.
(c) For two arbitrary faces $F, G \in \mathcal{F}(P)$ there exists a uniquely determined largest face $F \wedge G$ such that $F \supseteq F \wedge G$ and $G \supseteq F \wedge G$.

The properties described in Exercise 3.14 show that $(\mathcal{F}(P), \subseteq)$ is a *lattice*, called the *face lattice* of P.

Definition 3.15 A *combinatorial isomorphism* of two polytopes is a (poset-)isomorphism of the face lattices. If there exists such a combinatorial isomorphism, we call the two polytopes *combinatorially equivalent*. The *combinatorial type* of a polytope is the isomorphism type of its face lattice.

Exercise 3.16 Show that every affine transformation of a polytope P to a polytope Q induces an isomorphism from $\mathcal{F}(P)$ to $\mathcal{F}(Q)$.

Exercise 3.17 Give an example of two combinatorially equivalent polytopes such that there does not exist an affine transformation that maps one to the other.

Theorem 3.18 *Let F and G be faces of P such that $F \subseteq G$. Then*

$$\mathcal{F}(F, G) := \left\{ F' \in \mathcal{F}(P) : F \subseteq F' \subseteq G \right\}$$

with the partial order induced by containment, is isomorphic to the face lattice of a polytope of dimension $\dim G - \dim F - 1$.

Fig. 3.5 Two examples of polytopes $P \cap H(\epsilon)$ for $\dim F \in \{0, 1\}$. (Notation as in the proof of Theorem 3.18.) The polytope P is a *bipyramid* over a pentagon

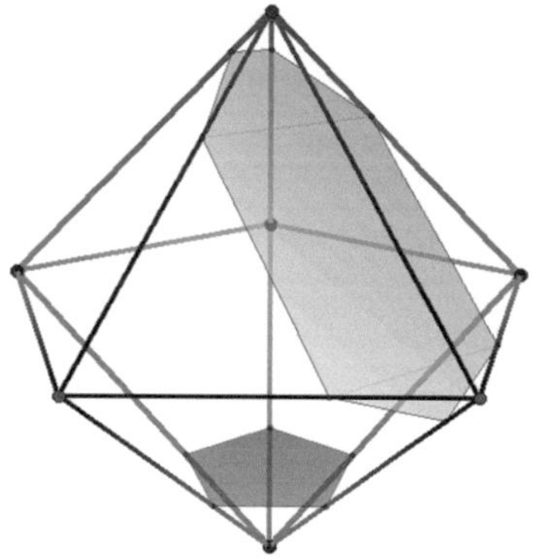

Proof Since Theorem 3.6 established that every face of a polytope is a polytope itself, we can assume without loss of generality that $G = P$. Let P be full-dimensional. We choose F as a proper face since otherwise there is nothing left to show.

Let V be the vertex set of P and $V(F) = F \cap V$ the vertex set of the face F. Choose a supporting hyperplane H to P with $F = P \cap H$. We assume that H has the oriented homogeneous coordinates $[a_0 : \cdots : a_n]$ and that $P \subseteq H^+$ holds. For every sufficiently small $\epsilon > 0$ we have that the hyperplane $H(\epsilon) = [a_0 - \epsilon : a_1 : \cdots : a_n]$, which is parallel to H, separates the vertex set $V(F)$ from its complement: $V(F) \subseteq \operatorname{int} H(\epsilon)^-$ and $V \setminus V(F) \subseteq \operatorname{int} H(\epsilon)^+$. See Fig. 3.5.

Let x be a point in the relative interior of F. The hyperplane $H(\epsilon)$ contains an interior point, y say, of P. Now let A be an $(n - \dim F)$-dimensional affine subspace containing x and y but no point in $\operatorname{aff} F$ other than x. That is, $\operatorname{aff} F$ and A are complementary affine subspaces meeting at x. Then $A \cap H(\epsilon)$ is an affine subspace of dimension $n - \dim F - 1$ which is affinely generated by the set

$$P(F, A, \epsilon) := P \cap A \cap H(\epsilon),$$

which by Theorem 3.10 is a polytope. The map

$$\alpha : \mathcal{F}(F, P) \to \mathcal{F}\big(P(F, A, \epsilon)\big) : F' \mapsto F' \cap A \cap H(\epsilon)$$

respects containment and is bijective since $\alpha^{-1}(F' \cap A \cap H(\epsilon)) = \operatorname{aff}((F' \cap A \cap H(\epsilon)) \cup F) \cap P = \operatorname{aff}(F') \cap P = F'$. Since $\mathcal{F}(F, P)$ does not depend on x, y, A or ϵ we have that the combinatorial type of the polytope $P(F, A, \epsilon)$ is independent of x, y, A and ϵ. $\qquad\square$

Given a face F of P we call the polytope $P(F, A, \epsilon)$ a *face figure* of F. An implication of Theorem 3.18 is the fact that $\mathcal{F}(F, G)$, for $\dim G - \dim F = 2$, is always the face lattice of a segment, i.e., there exist exactly two faces E_1 and E_2 of dimension $\dim F + 1$ which lie between F and G. This property is called the *diamond property* of $\mathcal{F}(P)$; see Fig. 3.6. For a bipyramid over a pentagon, Fig. 3.5 shows the (pentagonal) face figure of a vertex, also called a *vertex figure*; the shaded

Fig. 3.6 The diamond
property of the face lattice.
The faces E_1, E_2, F, G
satisfy $E_1 \vee E_2 = G$ and
$E_1 \wedge E_2 = F$

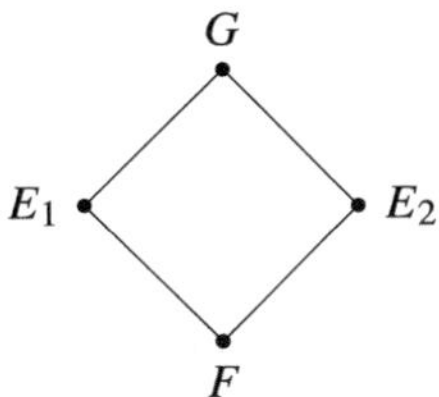

heptagon corresponds to the face figure of an edge, which is a segment, as discussed
above.

Let $P \subseteq \mathbb{R}^n$ be an n-polytope and let $f_k(P)$ be the number of k-dimensional faces
of P. Then $f(P) := (f_0(P), f_1(P), \ldots, f_{n-1}(P))$ is called the *f-vector* of P. The
f-vector is clearly a combinatorial invariant since it only depends on the combinatorial type of P. An interesting—and very complicated—task is to determine which
n-tuples of natural numbers can be f-vectors of n-polytopes.

Exercise 3.19 Compute the f-vector of the n-dimensional standard cube C_n and
describe its face lattice.

One may ask what "typical" polytopes look like. A more rigorous statement of
this naive question can be formulated in several ways using stochastic terms. As
an example, we will study the convex hulls of random points on the unit sphere in
Section 3.6. In many cases the term "typical" corresponds to "general position".

Exercise 3.20 Let $K \subseteq \mathbb{R}^n$ be a full-dimensional convex set. Show that a finite set
X of uniformly distributed random points from K is almost certainly in *general
position*, i.e. the probability of $n + 1$ of these points being affinely independent is 1.
In particular this implies that every proper face of conv X is a simplex.

The last property inspires the following definition.

Definition 3.21 A polytope P is called *simplicial* if all proper faces of P are simplices. It is called *simple* if the face figure of every proper face of P is a simplex.

The cross polytopes conv$\{\pm e^{(1)}, \ldots, \pm e^{(n)}\}$ are simplicial, while the cubes C_n
are simple. The relationship between these two properties, simplicial and simple,
will be clarified in Section 3.3.

Exercise 3.22 Show that a polytope is both simplicial and simple if and only if it is
a simplex or a polygon.

Exercise 3.23 Let P be an n-polytope with vertex set V and edge set E. The *graph*
$\Gamma(P)$ is the abstract graph (V, E) with natural incidence. Show:

(a) The graph $\Gamma(P)$ is connected.
(b) Every vertex is incident with at least n edges.

(c) The n-polytope P is simple if and only if every vertex is incident with exactly
   n edges.

3.3 Polarity and Duality

In the following section we introduce the concept of polarity. Given a polytope P
which contains the origin in its interior, we assign to P a polar polytope P° such
that every k-face of P corresponds to an $(n - k - 1)$-face of P°. In particular, we
have that $f_{n-i-1}(P) = f_i(P^\circ)$.

Example 3.24 For the standard cube $C_3 = [-1, 1]^3$ in $\mathbb{R}^3$ we have $f_0(C_3) = 8$,
$f_1(C_3) = 12$, $f_2(C_3) = 6$. For the three-dimensional cross-polytope (the octahe-
dron) $Q = \mathrm{conv}\{\pm e^{(i)} : 1 \leq i \leq 3\}$ (where $e^{(i)}$ denotes the i-th standard basis vector), we have $f_0(Q) = 6$, $f_1(Q) = 12$, $f_2(Q) = 8$ (see Fig. 3.7).
   We will see in Example 3.30 that Q is the polar polytope of C_3.

   As in Section 2.3.1, let $\langle \cdot, \cdot \rangle$ denote the Euclidean scalar product and let $\| \cdot \|$ with
$\|x\| := \langle x, x \rangle^{1/2}$ be the Euclidean norm.

Definition 3.25 For $X \subseteq \mathbb{R}^n$ the *polar set* X° is defined as

$$X^\circ = \{ y \in \mathbb{R}^n : \langle x, y \rangle \leq 1 \text{ for all } x \in X \}.$$

Exercise 3.26 Show that $X \subseteq Y$ implies $Y^\circ \subseteq X^\circ$ for $X, Y \subseteq \mathbb{R}^n$.

Proposition 3.27 *Let $X \subseteq \mathbb{R}^n$. Then X° is closed and convex and $0 \in X^\circ$.*

Proof Clearly $0 \in X^\circ$. Let $x \in \mathbb{R}^n \setminus \{0\}$, then

$$\{x\}^\circ = \{ y \in \mathbb{R}^n : \langle x, y \rangle \leq 1 \} = [1 : -x_1 : \cdots : -x_n]^+$$

is a closed affine half-space and $\{0\}^\circ = \mathbb{R}^n$. The intersection $X^\circ = \bigcap_{x \in X}\{x\}^\circ$ of
closed and convex sets is again closed and convex. $\square$

Theorem 3.28 *If $P \subseteq \mathbb{R}^n$ is an n-polytope with $0 \in \mathrm{int}\, P$, then P° is also an n-
polytope with $0 \in \mathrm{int}\, P$. We have*

$$P^\circ = \bigcap_{v \in V} \{ y \in \mathbb{R}^n : \langle v, y \rangle \leq 1 \} = \bigcap_{v \in V} [1 : -v_1 : \cdots : -v_n]^+, \tag{3.1}$$

where V is the vertex set of P.

Proof Since P is bounded, we have that P is contained in an open ball $B(0, \rho)$
with center 0 and radius ρ. For all $x \in \mathbb{R}^n$ with $\|x\| \leq 1/\rho$ the Cauchy–Schwarz

Fig. 3.7 The cube $[-1, 1]^3$
and octahedron
$\text{conv}\{e^{(i)} : 1 \le i \le 3\}$

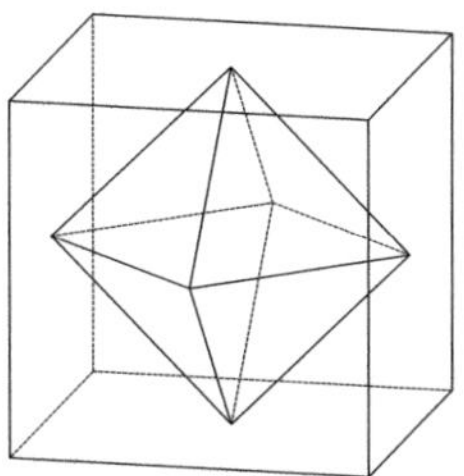

inequality gives

$$\left| \langle x, y \rangle \right| \le \|x\| \, \|y\| \le \frac{1}{\rho} \|y\| \le 1 \quad \text{for all } y \in P,$$

and thus the ball $B(0, 1/\rho)$ is contained in P°. As a consequence, P° is full-dimensional. Since P contains a ball $B(0, \rho')$ we can analogously deduce that P° is bounded.

Equation (3.1) remains to be proven. The inclusion "$\subseteq$" follows immediately from Definition 3.25. For the reverse inclusion "$\supseteq$" consider a point y that is not contained in P°. An arbitrary point $x \in P$ can be expressed as a convex combination $\sum_{i=1}^{k} \lambda^{(i)} v^{(i)}$ of vertices of P. Clearly we have

$$\langle x, y \rangle = \sum_{i=1}^{k} \lambda^{(i)} \langle v^{(i)}, y \rangle \le \max\{ \langle v^{(i)}, y \rangle : 1 \le i \le k \},$$

where the last inequality follows from $\sum_{i=1}^{k} \lambda^{(i)} = 1$. If $\langle x, y \rangle > 1$ then there exists a vertex $v^{(i)}$ such that $\langle v^{(i)}, y \rangle > 1$, which proves the statement. $\square$

Theorem 3.29 *For an n-polytope $P \subseteq \mathbb{R}^n$ with $0 \in \text{int } P$ we have*:

(a) $(P^\circ)^\circ = P$.

(b) *For every boundary point p of P the affine hyperplane*

$$H = \{ x \in \mathbb{R}^n : \langle p, x \rangle = 1 \}$$

is a supporting hyperplane to P°.

Proof (a) The definition of polarity implies immediately that $P \subseteq (P^\circ)^\circ$. For the reverse direction let $P = \bigcap_{i=1}^{m} H_i^+$ and let x be a point not contained in P. Then there exists an $i \in \{1, \ldots, m\}$ with $x \notin H_i^+$. By the Separation Theorem 2.14 there exists a $v \in \mathbb{R}^n$ with $\langle v, x \rangle > 1$. Since $\langle v, y \rangle \le 1$ for all $y \in H_i^+$, we have $v \in P^\circ$ and since $\langle v, x \rangle > 1$ we have $x \notin (P^\circ)^\circ$.

(b) For every $p \in P \setminus \{0\}$, $H^+ = [1 : -p_1 : \cdots : -p_n]^+$ is a half-space which contains the polytope P°. If p is a boundary point of P then, by Theorem 3.7, it belongs to a face of P and there exists a vector $x \in \mathbb{R}^n$ such that the hyperplane $H' = \{ y \in \mathbb{R}^n : \langle x, y \rangle = 1 \}$ supports P and contains the point p. We now have

$x \in P^\circ$ and $x \in H$ such that H intersects the polytope P° and thus H is a supporting hyperplane to P°.   □

Example 3.30 The description of the hyperplanes that define the facets of the standard cube C_n in Example 3.11 shows that C_n is polar to the n-dimensional cross-polytope.

Lemma 3.31 *Let $P \subseteq \mathbb{R}^n$ be an n-polytope with $0 \in \operatorname{int} P$. For every proper face F of P the face*

$$F^* := \big\{ x \in P^\circ : \langle x, y \rangle = 1 \text{ for all } y \in F \big\} \subsetneq F^\circ$$

is a proper face of P°.

Proof For every $p \in F$, Theorem 3.29b implies that the hyperplane $H = \{x \in \mathbb{R}^n : \langle p, x \rangle = 1\}$ is a supporting hyperplane of P°, and thus $P^\circ \cap H$ is a face of P°. Since F is the convex hull of a finite number of points, we can express F^* as an intersection of a finite number of such faces of P°. Hence F^* is a face of P°. By construction the face F^* is contained in the set F°.   □

Lemma 3.31 induces a map $\phi : F \mapsto F^*$ from the set of all proper faces of P to the set of all proper faces of P°, with $\phi(P) = \emptyset$ and $\phi(\emptyset) = P^\circ$.

Theorem 3.32 *Let $P \subseteq \mathbb{R}^n$ be an n-polytope with $0 \in \operatorname{int} P$. The map ϕ is bijective and for all $k \in \{0, \dots, n-1\}$ it maps the k-faces of P to the $(n-k-1)$-faces of P°. Furthermore ϕ is containment-reversing, i.e., $F \subseteq G$ implies $G^* \subseteq F^*$.*

Proof Exercise 3.26 implies that ϕ is containment-reversing.

Since the map ϕ can also be applied to the faces of P°, in order to prove bijectivity we know by Theorem 3.29a that it suffices to show that $\phi(\phi(F)) = F$ for every face F of P. For the non-proper faces this is satisfied by definition. For every proper face F we have by definition that

$$\phi\big(\phi(F)\big) = \big\{ x \in \mathbb{R}^n : \langle x, y \rangle = 1 \text{ for all } y \in \phi(F) \big\}$$

with

$$\phi(F) = \big\{ y \in P^\circ : \langle x, y \rangle = 1 \text{ for all } x \in F \big\},$$

and thus $F \subseteq \phi(\phi(F))$.

For the reverse inclusion, consider a point $p \in P$ with $p \notin F$. If we denote by $H = [1 : -h_1 : \cdots : -h_n]$ a supporting hyperplane to P that contains F, then $p \in H_\circ^+$ and hence $\langle p, h \rangle < 1$. Notice the first coordinate of H equals 1 (up to scaling by a positive real number) as the origin is an interior point of P. Since $h \in \phi(F)$ it then follows that $p \notin \phi(\phi(F))$.

Our dimension statement remains to be proven. For non-proper faces it is clearly satisfied. Every proper k-face F contains $k+1$ affinely independent points such

Fig. 3.8 A polytope $P = \mathrm{conv}\{p^{(1)}, \ldots, p^{(4)}\}$ and its polar polytope

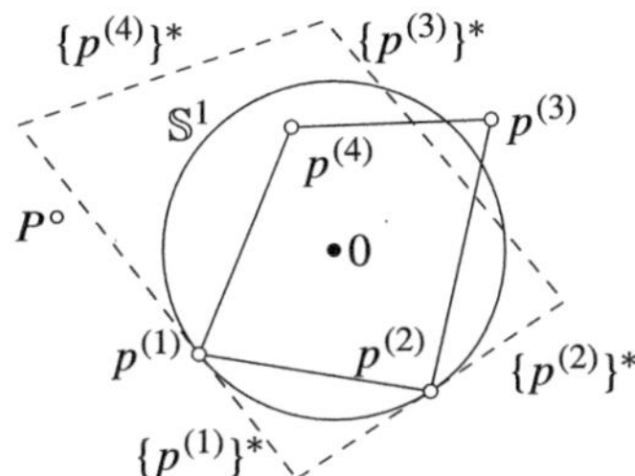

that $\phi(F)$ is contained in the intersection of $k+1$ hyperplanes whose equations are linearly independent. This implies $\dim \phi(F) \leq n - k - 1$, but since $F \subseteq \phi(\phi(F))$, this must be an equality. $\qquad\square$

A bijection that reverses the order relation of a lattice or poset is called an *anti-isomorphism*.

Corollary 3.33 *Let $P \subseteq \mathbb{R}^n$ be an n-polytope with $0 \in \mathrm{int}\, P$. The polytope P is simplicial if and only if the polar polytope P° is simple.*

Example 3.34 Let $P = \mathrm{conv}\{p^{(1)}, \ldots, p^{(4)}\} \subseteq \mathbb{R}^2$ be the quadrangle depicted in Fig. 3.8. Then the polar polytope $P^\circ = \bigcap_{i=1}^{4}\{x \in \mathbb{R}^2 : \langle p^{(i)}, x \rangle \leq 1\}$ is the dashed quadrangle. The segment $[0, p^{(i)}]$ is perpendicular to the line

$$H_i := \left\{x \in \mathbb{R}^2 : \langle p^{(i)}, x \rangle = 1\right\} = \mathrm{aff}\left(\{p^{(i)}\}^*\right).$$

For $i \in \{1, 2\}$, $p^{(i)}$ lies on the unit circle $\mathbb{S}^1$ such that the line H_i is tangent to the unit circle. The point $p^{(3)}$ lies outside of the unit circle such that H_3 intersects the interior of the unit circle. The fourth point $p^{(4)}$ lies in the interior of the unit circle such that H_4 lies completely outside of the unit circle. The distance from the origin to the line H_i is always the reciprocal of the distance between 0 and $p^{(i)}$.

In this section we often assumed that the polytope P contained the origin as an interior point. By restricting to $\mathrm{aff}\, P$, and via a suitable translation, we can without any loss of generality always assume this to be the case, i.e., every polytope has an affine image which satisfies this condition. This implies that for every polytope P there exists a polytope P' whose face lattice $\mathcal{F}(P')$ is anti-isomorphic to $\mathcal{F}(P)$. Such a polytope is said to be *dual* to P.

3.4 Polyhedra

Polytopes are the elementary building blocks of computational geometry, but it is often more natural to study a wider class of objects: polyhedra. This will be of particular importance in Chapter 4, as well as other chapters, when we discuss linear programming.

Definition 3.35 A set $P \subseteq \mathbb{R}^n$ is called a *polyhedron* if it can be represented as the intersection of a finite number of closed affine half-spaces.

Thus, a polytope is a bounded polyhedron. In general we cannot describe a polyhedron as the convex hull of a finite number of points. The most basic example of this is a single half-space. Nevertheless, the differences between polytopes and unbounded polyhedra are manageable.

To do this we distinguish between two kinds of unbounded polyhedra: A polyhedron P either contains an affine line or it does not. In the latter case we call P *pointed*. First assume that $P = H_1^+ \cap \cdots \cap H_k^+$ is pointed. Then $n \leq k$ and $H_1 \cap \cdots \cap H_k$ is either empty or contains exactly one point. Without loss of generality we may assume that the first n hyperplanes intersect in a point, z. If H_i^+ has homogeneous coordinates $[h_0^{(i)} : \cdots : h_n^{(i)}]^+$ this means that

$$z = \left[h_0^{(1)} : \cdots : h_n^{(1)}\right] \cap \cdots \cap \left[h_0^{(n)} : \cdots : h_n^{(n)}\right].$$

The point z may or may not be contained in P. To get a clearer image of P we apply an affine transformation to P which transforms each hyperplane H_i to the coordinate hyperplane $E_i = [0 : \cdots : 0 : 1 : 0 : \cdots : 0] = \{x \in \mathbb{R}^n : x_i = 0\}$ for $1 \leq i \leq n$. In this way, z will automatically be mapped to the origin. The affine transformation described above can be represented most conveniently by the $(n+1) \times (n+1)$-matrix

$$T = \begin{pmatrix} 1 & 0 & \cdots & 0 \\ h_0^{(1)} & h_1^{(1)} & \cdots & h_n^{(1)} \\ \vdots & \vdots & \ddots & \vdots \\ h_0^{(n)} & h_1^{(n)} & \cdots & h_n^{(n)} \end{pmatrix},$$

which operates on the left, as usual. That T indeed has all the properties required is a consequence of Exercise 2.17. By our choice of coordinates, $E_1, \ldots, E_n$ are oriented so that $E_1^+ \cap \cdots \cap E_n^+$ is the positive orthant. Hence the transformed polyhedron $[T]P$ is contained in the positive orthant. Now consider a further projective transformation, defined by the non-negative matrix

$$B = \begin{pmatrix} 1 & 1 & 1 & \cdots & 1 & 1 \\ 0 & 1 & 0 & \cdots & 0 & 0 \\ 0 & 0 & 1 & \cdots & 0 & 0 \\ \vdots & \vdots & & \ddots & & \vdots \\ 0 & 0 & 0 & \cdots & 1 & 0 \\ 0 & 0 & 0 & \cdots & 0 & 1 \end{pmatrix}.$$

The map $[B]$ is not an affine transformation. It maps the ideal hyperplane $[1 : 0 : \cdots : 0]$ to the projective hyperplane $[1 : -1 : \cdots : -1]$ so that the coordinate hyperplanes stay fixed. Furthermore the image of the positive orthant under the map $[B]$ is the n-simplex

$$E_1^+ \cap \cdots \cap E_n^+ \cap [1 : -1 : \cdots : -1]^+.$$

In particular, the image $[BT]P$ is a bounded polyhedron, i.e., a polytope. We have now proved the following theorem.

Theorem 3.36 *Every pointed polyhedron is projectively equivalent to a polytope.*

Pointed polyhedra can be imagined as polytopes with a specific proper face that has been moved to the ideal hyperplane. For an example computation, see Section 3.6.3 below.

Note that the image of a polyhedron under a projective transformation is not necessarily a polyhedron. However, the following case, which is most relevant for us, does not have this problem.

Exercise 3.37 Let $P \subseteq \mathbb{R}^n_{\geq 0}$ be a polyhedron in the positive orthant and $[A]$ the projective transformation to a matrix $A \in \mathrm{GL}_{n+1}\,\mathbb{R}$ with non-negative coefficients. Show that the image $[A]P$ is again a polyhedron.

We still need to consider the case where P is not pointed. In this case we choose an affine subspace A of $\mathbb{R}^n$ which is contained in P and which is maximal with respect to dimension. The linear subspace L of $\mathbb{R}^n$ which is parallel to A is called the *lineality space* of P. Let p be an arbitrary point of P and A' be the affine orthogonal complement of A that contains p. The intersection $P \cap A'$ is a polyhedron which contains no affine line and is therefore pointed.

Definition 3.38 For $X, Y \subseteq \mathbb{R}^n$, the *Minkowski sum* of X and Y is defined as

$$x + y = \{x + y : x \in X, \, y \in Y\}.$$

The Minkowski sum is called *direct* if $x + y = v + w$ with $x, v \in X$ and $y, w \in Y$ implies $x = v$ and $y = w$.

Using this notation, the lineality space, as defined above, gives us a direct Minkowski sum $P = (P \cap A') + L$, establishing the following lemma.

Lemma 3.39 *Every polyhedron can be expressed as the direct Minkowski sum of a pointed polyhedron and a linear subspace.*

In this decomposition it is possible that the pointed polyhedron or the lineality space is just a single point. In those cases the decomposition as a Minkowski sum is trivial.

In summary, we can say that statements about polyhedra can be traced back to statements about polytopes. As an example of this consider the generalization of Theorem 3.8 which is discussed in Exercise 3.41. However, first consider two further definitions.

Definition 3.40 Let $A \subseteq \mathbb{R}^n$. A *positive combination* of A is a linear combination $\sum_{i=1}^{m} \lambda^{(i)} a^{(i)}$ with $a^{(i)} \in A$ and $\lambda^{(i)} \geq 0$ for all i. The set of all positive combinations of A is called the *positive hull* of A, which we denote by $\mathrm{pos}\,A$.

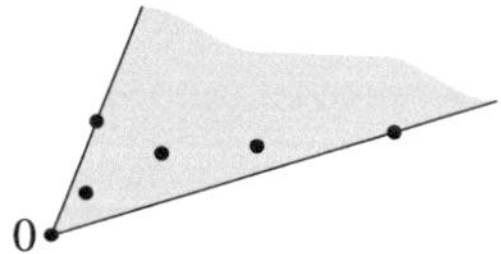

Fig. 3.9 The positive hull of
a finite point set

The positive hull of a set A is a *convex cone* in the sense that pos A is convex and
that $a + b$ and λa are contained in pos A for $a, b \in$ pos A and $\lambda \geq 0$ (see Fig. 3.9).
In a convex cone K we call a half-line $x + \mathbb{R}_{\geq 0} y \subseteq \partial K$ with $x \in K$ and $y \in \mathbb{R}^n \setminus \{0\}$
a *ray* of K.

Exercise 3.41 Every polyhedron $P \subseteq \mathbb{R}^n$ can be expressed as a Minkowski sum

$$P = \mathrm{conv}\ V + \mathrm{pos}\ R$$

for finite sets V and R.

Exercise 3.42 Show that the cone pos R in Exercise 3.41 is uniquely determined. Is
the polytope conv V also unique, in general?

The cone pos R in the preceding exercises is called the *recession cone* of the
polyhedron P.

Exercise 3.43 The *product*

$$\left\{ (p, q) \in \mathbb{R}^{n+n'} : p \in P,\ q \in Q \right\}$$

of two polyhedra $P \subseteq \mathbb{R}^n$ and $Q \subseteq \mathbb{R}^{n'}$ is a polyhedron.

3.5 The Combinatorics of Polytopes

As mentioned in Chapter 1, computational problems often require the transforma-
tion from a $\mathcal{V}$-representation to an $\mathcal{H}$-representation and vice versa. Before we study
explicit algorithms for this task in Chapter 5, it is necessary to improve our understanding of the combinatorial structure of polytopes.

To be able to discuss the complexity of an algorithm it is necessary to first determine how large the output of an algorithm may be in relation to its input. For convex-
hull-algorithms, i.e., methods to compute the facets of a convex hull of a given point
set, we have to answer the question of how many facets an n-dimensional polytope
with m vertices may have. The reverse question, how many vertices an n-polytope
with m facets can have, is equivalent by polarization. As before, let $f_k(P)$ denote
the number of k-dimensional faces of an n-polytope P for $-1 \leq k \leq n$. In particular
we have $f_{-1}(P) = f_n(P) = 1$.

The Upper-bound Theorem, a fundamental result in polytope theory, states that
the cyclic polytopes from Definition 3.2 are extremal in the following sense. Let

$Z_n(m)$ denote a cyclic polytope in $\mathbb{R}^n$ formed by the convex hull of m points on the moment curve. This notation purposefully neglects which m points define the cyclic polytope. At the end of this section this simplification will be justified (in Exercises 3.49–3.51) by the statement that two such cyclic polytopes are combinatorially equivalent.

Theorem 3.44 (Upper-bound Theorem, McMullen 1970) *An n-dimensional polytope with m vertices has at most as many k-faces as a cyclic polytope $Z_n(m)$ for all $k \in \{-1, \ldots, n\}$.*

In Exercises 3.49–3.51 we will compute the number of facets of cyclic polytopes. This yields the following explicit upper bound.

Corollary 3.45 *The number of facets of an n-dimensional polytope with m vertices is bounded by*

$$\begin{cases} \frac{m}{m-\frac{n}{2}}\binom{m-\frac{n}{2}}{m-n} & \text{if } n \text{ is even,} \\[2ex] 2\binom{m-\frac{n+1}{2}}{m-n} & \text{if } n \text{ is odd.} \end{cases}$$

When we switch to the dual picture we obtain the same upper bound for the number of vertices of an n-polytope with m facets.

We will not fully prove Theorem 3.44 in this section. Instead we give a proof for an upper bound which has the right order of magnitude for the number of facets.

Theorem 3.46 *An n-polytope with m vertices has at most $2\binom{m}{\lfloor n/2 \rfloor}$ facets and in total not more than $2^{n+1}\binom{m}{\lfloor n/2 \rfloor}$ faces. For fixed n, both numbers have the same order of magnitude $O(m^{\lfloor n/2 \rfloor})$.*

We will prove this statement first for simplicial polytopes and then we deduce the non-simplicial case as a corollary.

Lemma 3.47 *For a simplicial n-polytope P we have:*

(a) $(n-k)f_k(P) \le \binom{n}{k+1}f_{n-1}(P)$ *for* $k \in \{-1, \ldots, n\}$;
(b) $nf_0(P) + (n-1)f_1(P) + \cdots + 2f_{n-2}(P) \le (2^n - 2)f_{n-1}(P)$;
(c) $f_{n-1}(P) \le 2f_{\lfloor n/2 \rfloor - 1}(P)$.

Proof For the first statement we count the number of k-faces that are incident to a given facet of P and vice versa. By our assumption every facet is an $(n-1)$-simplex which contains exactly $\binom{n}{k+1}$ k-faces. On the other hand we have that the face figure of a k-face is an $(n-k-1)$-polytope that has at least $n-k$ facets. This implies the first statement. The second statement follows from the first by summation over k from 0 to $n-2$.

For the third statement, we consider the dual polytope P' which is by Corollary 3.33 simple. We have to show that $f_0(P') \le 2f_{\lceil n/2 \rceil}(P')$.

Now we will limit the number of vertices of P' with respect to the number of $\lceil n/2 \rceil$-faces. After an affine transformation we can assume without loss of generality that no two edges of P' have the same x_n-coordinate. In the following we imagine that the n-coordinate is "pointing upwards".

Consider a vertex v and the n edges incident with v. Then there are at least $\lceil n/2 \rceil$ edges that point downwards or at least $\lceil n/2 \rceil$ edges that point upwards. In the first case we have that every $\lceil n/2 \rceil$-tuple of upward pointing edges determines a $\lceil n/2 \rceil$-face for which v is the lowest vertex. In the second case we have that each $\lceil n/2 \rceil$-tuple of downward pointing edges determines a $\lceil n/2 \rceil$-face for which v is the highest vertex. Since the lowest and highest vertex for each face are unique, there are at most twice as many vertices as there are $\lceil n/2 \rceil$-faces. $\square$

Lemma 3.48 *For each n-polytope P there exists an n-dimensional simplicial polytope Q with the same number of vertices as P such that $f_k(Q) \geq f_k(P)$ for $1 \leq k \leq n$.*

Proof We can assume that $P \subseteq \mathbb{R}^n$. Our goal is to obtain the polytope Q from P by slightly moving all the vertices.

For the *perturbation* of one vertex v we employ the following operation. Pick an affine hyperplane H which strictly separates v from all the other vertices of P. We may orient H so that $H_\circ^-$ contains v. Now choose a point $v' \in \operatorname{int} P \cap H_\circ^-$ which is not contained in a hyperplane spanned by any $n + 1$ vertices of P. Replacing v by v' we obtain the polytope

$$P' = \operatorname{conv}\big(\{w : w \text{ vertex of } P \text{ distinct from } v\} \cup \{v'\}\big),$$

which is contained in P. We want to show that P' has at least as many faces of each dimension as P. To this end we will describe an injective map ι from the faces of P to the faces of P'.

Let F be a proper face of P and let A be an affine hyperplane supporting P with $A \cap P = F$. Notice that, as P contains P', the hyperplane A does not separate P'. If F does not contain the vertex v then we set $\iota(F) = A \cap P' = A \cap P = F$, and this is a face of P'.

It remains to consider the case when F contains v. If F is a simplex, then

$$\iota(F) = \operatorname{conv}\big(\{w : w \text{ vertex of } F \text{ distinct from } v\} \cup \{v'\}\big)$$

is a face of P'. We may then assume that F is not a simplex. As v' is not contained in F in this case it follows that $\iota(F) = A \cap P'$ is a face of P' of the same dimension as F. This yields a dimension-preserving map ι from the face lattice of P to the face lattice of P'. It is easy to see that ι is injective.

To construct the polytope Q we pick a linear ordering $v^{(1)}, \ldots, v^{(m)}$ of the vertices of P. Inductively perturbing the vertices in this order gives a sequence of n-dimensional polytopes $P^{(1)}, \ldots, P^{(m)}$ all of which have precisely m vertices. Setting $P^{(0)} = P$ and $Q = P^{(m)}$ we have

$$f_k(P) \leq f_k\big(P^{(i)}\big) \leq f_k(Q)$$

for $1 \leq i \leq m$ and $1 \leq k \leq n$. Moreover, our procedure guarantees that the vertices of Q are in general position and therefore Q is simplicial. $\square$

Proof of Theorem 3.46 By Lemma 3.48 it suffices to consider simplicial polytopes P. Since the number of $(\lfloor n/2 \rfloor - 1)$-faces clearly satisfies

$$f_{\lfloor n/2 \rfloor - 1}(P) \leq \binom{m}{\lfloor n/2 \rfloor},$$

Lemma 3.47c implies

$$f_{n-1}(P) \leq 2 \binom{m}{\lfloor n/2 \rfloor}$$

and using Lemma 3.47b we obtain

$$f_0(P) + f_1(P) + \cdots + f_n(P) \leq 2^{n+1} \binom{m}{\lfloor n/2 \rfloor}.$$ $\square$

To conclude this section we will study the cyclic polytopes introduced in Definition 3.2. As mentioned in Theorem 3.44, these polytopes maximize the f-vector of all polytopes.

Exercise 3.49 Show that each set of n points on the moment curve in $\mathbb{R}^n$ are affinely independent. This implies that cyclic polytopes are simplicial.

As a result of the following exercise (and Exercise 3.55) we know that two cyclic polytopes of the same dimension and the same number of vertices are combinatorially equivalent. This justifies the notation $Z_n(m)$.

Exercise 3.50 (Gale Evenness Condition) Let V be the vertex set of a cyclic polytope in $\mathbb{R}^n$ with the induced order $\prec$ with respect to the moment curve, i.e., $x(\tau_1) \prec x(\tau_2)$ if and only if $\tau_1 < \tau_2$. Let $U = \{v^{(1)}, \ldots, v^{(n)}\} \subseteq V$ be an n-tuple of vertices of P, where $v^{(1)} \prec v^{(2)} \prec \cdots \prec v^{(n)}$. Show that conv U is a facet of P if and only if for every two vertices $u, v \in V \setminus U$ we have that the number of vertices $v^{(i)} \in U$ with $u \prec v^{(i)} \prec v$ is even.

Exercise 3.51 Show, using the evenness criterion from the previous exercise, that the following holds for the number $f_{n,m}$ of facets of a cyclic polytope $Z_n(m)$:

$$f_{n,m} = \begin{cases} \dfrac{m}{m-\frac{n}{2}} \binom{m-\frac{n}{2}}{m-n} & \text{if } n \text{ is even,} \\[2ex] 2 \binom{m-\frac{n+1}{2}}{m-n} & \text{if } n \text{ is odd.} \end{cases} \tag{3.2}$$

Exercise 3.52 Compute the group of combinatorial automorphisms of each cyclic polytope.

In the remainder of this section we will discuss the relationship between the number of faces of varying dimensions of polytopes. These relations are essential for a deeper understanding of the combinatorics of polytopes (such as the proof of the exact statement of the Upper-bound Theorem).

The entries of the f-vector of a polytope are not independent of each other. This is easy to see for simple n-polytopes. Here, every vertex is incident with exactly n edges and conversely, every edge is incident with exactly two vertices. This implies $2f_1 = nf_0$. Since f_1 is an even number, this implies that every simple polytope of odd dimension has an even number of vertices. In the dual picture this means that each simplicial polytope of odd dimension has an even number of facets. Theorem 3.54 sharpens this statement. First we look at a famous result that holds for arbitrary polytopes.

Theorem 3.53 (Euler's formula)   *The f-vector of a non-empty polytope P of dimension n satisfies the following equation*

$$\sum_{k=-1}^{n} (-1)^k f_k(P) = 0.$$

Euler's formula implies that for two-dimensional polytopes the number of vertices and edges in a polygon is the same.

For three-dimensional polytopes we obtain the classical formula for the *Euler characteristic*:

$$f_0(P) - f_1(P) + f_2(P) = f_{-1}(P) + f_3(P) = 2. \tag{3.3}$$

Proof   We prove Euler's formula by induction over the dimension n of the polytope. For $n = 1$ each polytope has exactly two proper faces, i.e., its vertices, such that

$$\sum_{k=-1}^{1} (-1)^k f_k(P) = 1 - 2 + 1 = 0.$$

So let P be an n-polytope and let $m = f_0(P)$ be the number of vertices of P. After a suitable affine transformation we can assume without loss of generality that no two vertices have the same x_n-coordinates. Let $v^{(1)}, \ldots, v^{(m)}$ be the vertex set of P, ordered increasingly by their x_n-coordinate. Furthermore, let $H_1, \ldots, H_{2m-1}$ be horizontal (i.e., orthogonal to the x_n-axis) affine hyperplanes such that $v^{(i)} \in H_{2i-1}$, $1 \leq i \leq m$, and such that $v^{(i)}$ is the only vertex that is located between H_{2i-2} and H_{2i}. For a face F of P we define

$$\chi_j(F) = \begin{cases} 1 & \text{if } H_j \cap \operatorname{relint} F \neq \emptyset, \\ 0 & \text{otherwise} \end{cases}$$

for $1 \leq j \leq 2m - 1$.

Now we fix a face F and denote by $v^{(l)}$ the vertex with minimal x_n-coordinate. Similarly $v^{(u)}$ is the vertex with maximal x_n-coordinate. The horizontal hyperplanes that intersect the interior of F lie strictly between the hyperplanes H_{2l-1} and H_{2u-1}. If $\dim F \geq 1$ then we have $l \neq u$ and the number of hyperplanes with even index that intersect relint F exceeds the number of hyperplanes with odd index that intersect relint F by one. That is,

$$\sum_{j=2}^{2m-2} (-1)^j \chi_j(F) = 1.$$

Summing this equation over the set $\mathcal{F}_k(P)$ of k-faces of P yields

$$f_k(P) = \sum_{F \in \mathcal{F}_k(P)} \sum_{j=2}^{2m-2} (-1)^j \chi_j(F).$$

The alternating sum over all $k \geq 1$ yields

$$\sum_{k=1}^{n} (-1)^k f_k(P) = \sum_{j=2}^{2m-2} (-1)^j \sum_{k=1}^{n} (-1)^k \sum_{F \in \mathcal{F}_k(P)} \chi_j(F). \tag{3.4}$$

For $2 \leq j \leq 2m - 2$, $P_j := P \cap H_j$ has dimension $n - 1$, so that by the induction hypothesis we have

$$\sum_{k=0}^{n-1} (-1)^k f_k(P_j) = 1. \tag{3.5}$$

We distinguish between two cases:

j *even:* Each $(k - 1)$-face of P_j is the intersection of a k-face of P with the hyperplane H_j such that

$$f_{k-1}(P_j) = \sum_{F \in \mathcal{F}_k(P)} \chi_j(F), \quad \text{for } 1 \leq k \leq n.$$

Substituting this into (3.5) yields

$$\sum_{k=1}^{n} (-1)^{k-1} \sum_{F \in \mathcal{F}_k(P)} \chi_j(F) = 1. \tag{3.6}$$

j *odd:* Each $(k - 1)$-face of P_j is the intersection of a k-face of P with H_j, with the exception of the vertex $v^{((j+1)/2)}$ which is contained in H_j. We therefore have

$$f_0(P_j) = 1 + \sum_{F \in \mathcal{F}_1(P)} \chi_j(F),$$

$$f_{k-1}(P_j) = \sum_{F \in \mathcal{F}_k(P)} \chi_j(F), \quad 2 \le k \le n.$$

In this case, substituting into (3.5) yields

$$\sum_{k=1}^{n} (-1)^{k-1} \sum_{F \in \mathcal{F}_k(P)} \chi_j(F) = 0. \tag{3.7}$$

Multiplying (3.6) and (3.7) by $(-1)^{j+1}$ and substituting into (3.4) yields

$$\sum_{k=-1}^{n} (-1)^k f_k(P) = -1 + m + \sum_{k=1}^{n} (-1)^k f_k(P)$$

$$= -1 + m + (m-1) \cdot (-1) + (m-2) \cdot 0 = 0. \qquad \square$$

From this we can deduce, with a clever summation, a previously mentioned, far-reaching generalization of the relation $2f_1 = nf_0$ for simple polytopes. For the proof we refer to Ziegler [99, §8.3].

Theorem 3.54 (Dehn–Sommerville equations) *The f-vector of a simple n-polytope P satisfies the linear equations*

$$\sum_{j=0}^{k} (-1)^j \binom{n-j}{n-k} f_j(P) = f_k(P), \quad for\ k \in \{0, \dots, n\}.$$

Through duality we obtain a corresponding statement for simplicial polytopes.

3.6 Inspection Using `polymake`

We want to study some concrete examples of polytopes, most of the characteristics of which can be obtained using the results discussed above. Here, and in the following, we use the software `polymake`, which is briefly introduced in Appendix D.1. The shell-based interface uses a dialect of Perl.

3.6.1 Cyclic Polytopes

We study cyclic 4-polytopes $Z_4(7)$ with 7 vertices. The `polymake` function `cyclic` generates cyclic polytopes which can then be further examined. Each line starting with "`polytope >`" contains one command.

```
polytope > $Z_4_7 = cyclic(4,7);

polytope > print $Z_4_7->VERTICES;
1 0 0 0 0
1 1 1 1 1
1 2 4 8 16
1 3 9 27 81
1 4 16 64 256
1 5 25 125 625
1 6 36 216 1296

polytope > print $Z_4_7->DIM;
4

polytope > print $Z_4_7->F_VECTOR;
7 21 28 14

polytope > print dense($Z_4_7->VERTICES_IN_FACETS);
1 1 0 0 1 1 0
0 1 1 0 1 1 0
0 0 1 1 1 1 0
0 1 1 1 1 0 0
1 1 0 1 1 0 0
1 1 1 1 0 0 0
1 1 1 0 0 0 1
1 0 1 1 0 0 1
1 0 0 1 1 0 1
0 0 0 1 1 1 1
1 0 0 0 1 1 1
0 0 1 1 0 1 1
0 1 1 0 0 1 1
1 1 0 0 0 1 1
```

VERTICES, DIM, F_VECTOR and VERTICES_IN_FACETS are examples of
properties of a polymake object. Each line of the property VERTICES contains
the oriented homogeneous coordinates of a vertex. The vertices are implicitly enu-
merated, starting with 0, but the order is not relevant here. The output of VER-
TICES_IN_FACETS refers to this enumeration: Every line of the output matrix
corresponds to a facet and every column to a vertex, where the order of the columns
corresponds to the order of the vertices in the VERTICES section. A 1 at position
(i, j) indicates that the i-th facet is incident with the j-th vertex. If the dense()
command is omitted in the polymake-command line, we get the entire list of vertices for each facet.

   In the dense output of VERTICES_IN_FACETS we can immediately verify
Gale's evenness criterion from Exercise 3.50: In each line we have an even number
of 1s between two 0s. The matrix with coefficients in $\{0, 1\}$ coded in the property
VERTICES_IN_FACETS is called the *incidence matrix* with respect to the given
order of vertices and facets.

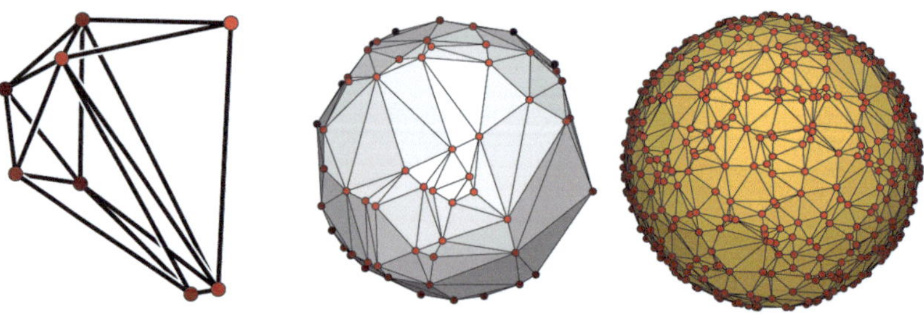

Fig. 3.10 The convex hulls of 8, 100 and 1000 random points on $\mathbb{S}^2$

The remaining two commands print the dimension and the f-vector of $Z_4(7)$. In particular, we see that this polytope has seven vertices, 21 edges, 28 ridges and 14 facets.

The homogeneous coordinates of the facets can be obtained using the command

```
polytope > print $Z_4_7->FACETS;
```

in the same order as for the property VERTICES_IN_FACETS. The output (which is suppressed here) looks similar to that in Section 5.4.

3.6.2 Random Polytopes

The function `rand_sphere` produces n-polytopes as convex hulls of random points selected from a uniform distribution on the unit sphere $\mathbb{S}^{n-1}$. By Exercise 3.20 these points are almost certainly in general position so that the convex hull is simplicial.

```
polytope > $R_3_8 = rand_sphere(3,8);
polytope > print $R_3_8->SIMPLICIAL;
1
polytope > print $R_3_8->F_VECTOR;
8 18 12
```

By the Dehn–Sommerville equations we know that the complete f-vector of a simplicial 3-polytope is determined by the number of vertices. We have $f_2 = 2f_0 - 4$ and $f_1 = f_0 + f_2 - 2 = 3f_0 - 6$. Figure 3.10 depicts some random polytopes.

3.6.3 Projective Transformations

We now want to show how to use `polymake` to projectively transform an unbounded but pointed polyhedron into a polytope. This is an example of the procedure described in Theorem 3.36. Let us begin by defining an unbounded polyhedron

as the Minkowski sum of a polytope (which is the convex hull of eight points) and
one infinite ray.

```
polytope > $P=new Polytope(POINTS=>
    [[0,0,0,1],
     [1,0,0,0],[1,3,0,0],[1,0,3,0],[1,3,3,0],
     [1,1,1,-1],[1,1,2,-1],[1,2,1,-1],[1,2,2,-1]]);
```

Again we use homogeneous coordinates, and the unique ray is represented by the
vector $(0, 0, 0, 1)^T$ listed first. It is easy to verify that $P satisfies our conditions.

```
polytope > print $P->BOUNDED, " ", $P->POINTED;
0 1
```

The first step is to set up an affine transformation which sends our polyhedron into
the positive orthant. To this end we list which facet is incident with which vertex.

```
polytope > print rows_numbered($P->FACETS_THRU_VERTICES);
0:1 2 5 6
1:0 1 2 3
2:0 1 6 8
3:2 3 5 7
4:5 6 7 8
5:0 3 4
6:3 4 7
7:0 4 8
8:4 7 8
```

In our input above the POINTS defining $P were, in fact, a non-redundant descrip-
tion. Hence the row numbers correspond to our input. For instance:

```
polytope > print $P->VERTICES->[5];
1 1 1 -1
```

We see that the facets numbered 0, 3 and 4 are incident with the vertex numbered 5.
Our polyhedron is full-dimensional and hence those three facets must have linearly
independent facet normal vectors. Therefore we can use vertex number 5 as our
point z in the construction described in Section 3.4. In this way we form the matrix
T as follows.

```
polytope > $T=new Vector([1,0,0,0])/$P
    ->FACETS->minor([0,3,4],All);
polytope > print $T;
1 0 0 0
0 0 1 1
0 1 0 1
1 0 0 1
```

Notice that the operator / concatenates matrices row-wise. Similarly, the matrix B
can be built from standard constructions using row and column concatenations, the
latter being expressed via $|$.

```
polytope > $B=ones_vector(4)/(zero_vector(3)|unit_matrix(3));
polytope > print $B;
1 1 1 1
0 1 0 0
0 0 1 0
0 0 0 1
```

To transform our polytope we have to take into account that `polymake` uses row vectors to represent the points, and thus transformations operate on the right. We can use the matrices constructed above if we transpose.

```
polytope > $Q=transform($P,transpose($B*$T));
polytope > print $Q->BOUNDED;
1
```

The vertices of the transformed polytope $Q are as follows.

```
polytope > print rows_numbered($Q->VERTICES);
0:1 1/3 1/3 1/3
1:1 0 0 1/2
2:1 0 3/5 1/5
3:1 3/5 0 1/5
4:1 3/8 3/8 1/8
5:1 0 0 0
6:1 1/2 0 0
7:1 0 1/2 0
8:1 1/3 1/3 0
```

Observe that the vertex numbered 5, which is the image of the point z under the transformation, is the origin. All the vertices are contained in the positive orthant.

3.7 Exercises

Exercise 3.55 Show that two polytopes are combinatorially equivalent if and only if there exists an ordering of their vertices and facets such that their corresponding incidence matrices are equal.

A polytope is called *cubical* if all of its proper faces are combinatorially equivalent to cubes. For cubical polytopes there is a statement which is analogous to Lemma 3.47. This is an observation of Gil Kalai.

Exercise 3.56 Show that the following inequality holds for the f-vector of a cubical polytope

$$f_1 + 2f_2 + 2^2 f_3 + \cdots + 2^{n-2} f_{n-1} \leq \binom{f_0}{2}.$$

Exercise 3.57 Show that, given an arbitrary full-dimensional polyhedron $P \subseteq \mathbb{R}^n$ with outer description $P = \bigcap_{i=1}^{m} H_i^+$, there exists a family of indices $i_0, i_1, \ldots, i_n$

such that $Q = H_{i_0}^+ \cap \cdots \cap H_{i_n}^+$ is projectively equivalent to an n-simplex. If we additionally assume that P is a polytope, can we always choose the hyperplanes in such a way that Q is also a polytope?

Exercise 3.58 Let $\pi : \mathbb{R}^{n+1} \to \mathbb{R}^n$ be the linear projection to the first n coordinates. Show that the image of a polytope under π is again a polytope.

Exercise 3.59 Let P be an n-polytope. Show that there exists for every k-face G of P a family of facets $F_1, \ldots, F_{n-k}$ such that

$$G_1 \supsetneqq G_2 \supsetneqq \cdots \supsetneqq G_{n-k} = G$$

holds for $G_i := F_1 \cap \cdots \cap F_i$.

Exercise 3.60 The Minkowski-sum $[p^{(1)}, q^{(1)}] + \cdots + [p^{(k)}, q^{(k)}]$ of a finite number of line segments with $p^{(i)}, q^{(i)} \in \mathbb{R}^n$ is a *zonotope*. Show that the zonotopes generated by k segments are exactly the images of the standard cube $[-1, 1]^k$ under affine maps.

Exercise 3.61 Which among the polyhedra in the following list are projectively equivalent?

(a) $\mathrm{conv}\{(0, 0, 0)^T\} + \mathrm{pos}\{(1, 0, 0)^T, (0, 1, 0)^T, (0, 0, 1)^T\}$
(b) $\mathrm{conv}\{(0, 0, 0)^T, (1, 0, 0)^T, (0, 1, 0)^T\} + \mathrm{pos}\{(1, 1, 1)^T\}$
(c) $\mathrm{conv}\{(1, 0, 0)^T, (0, 1, 0)^T, (0, 0, 1)^T, (a, b, c)^T\}$ for a, b, c arbitrary real numbers (i.e., this is an infinite set of polyhedra)
(d) $\mathrm{conv}\{(1, 0, 0)^T, (0, 1, 0)^T, (0, 0, 1)^T, (2, 1, 1)^T, (1, 2, 1)^T, (1, 1, 2)^T\}$

3.8 Remarks

The content of this chapter forms part of the standard material for polytopes and polyhedra, see the monographs of Boissonnat and Yvinec [15], Brøndsted [16], Grünbaum [56] and Ziegler [99]. The Upper-bound Theorem was proved by Mc-Mullen [78]. Further proofs can be found in the books of Mulmuley [80] and Ziegler [99]. As a general reference we recommend the *Handbook of Discrete and Computational Geometry* [49].

The term *polyhedron* is not always used in the same way as we have defined it in this book. In particular topologists often use this term to describe a simplicial or polyhedral complex embedded in $\mathbb{R}^n$; it may also describe a triangulated manifold.

The set of vertices and edges of a 3-dimensional polytope can be interpreted as the set of vertices and edges of a planar graph on a sphere. Hence (3.3) is a special case of Euler's formula for planar graphs, see [2, Chapter 11]. In fact, Euler's formula generalizes to cell complexes. We will see a glimpse of this in Section 13.1 at the very end of this book.

Concerning the graph $\Gamma(P)$ of an n-polytope P introduced in Exercise 3.23, Warren M. Hirsch conjectured in 1957 that any two vertices of P can be connected by a path in $\Gamma(P)$ of at most $m - n$ edges, where m is the number of facets. This famous conjecture, the *Hirsch conjecture*, was disproved by Santos in 2010 [90].

Chapter 4
Linear Programming

Many algorithms in computational geometry are based on methods from linear programming. The task of linear programming is to maximize or minimize a linear objective function on a polyhedron P which is given by inequalities. If P is non-empty and bounded, we will see that the optimal value is always attained at a vertex of P.

As in most textbooks on linear programming, the algorithms we present operate on matrices and their rows and columns. These matrices correspond directly to data structures which can be used to implement the algorithms. To strengthen our understanding of the geometric setting, we will reformulate our methods in the language of polytope theory.

4.1 The Task

In the following, for two vectors $x, y \in \mathbb{R}^n$, let $x \leq y$ denote the component-wise inequality relation. Furthermore, let $(\mathbb{R}^n)^*$ denote the dual space to $\mathbb{R}^n$, i.e., the vector space of all linear mappings $\mathbb{R}^n \to \mathbb{R}$. We study *linear programs*, abbreviated *LP*, of the form

$$\max\{cx : Ax \leq b\} \tag{4.1}$$

for a given $m \times n$-matrix A and vectors $b \in \mathbb{R}^m$ and $c \in (\mathbb{R}^n)^*$. Here, the right hand side b of the constraints is a column vector and the *objective function* c can be identified with a row vector.

The polyhedron

$$P(A, b) := \left\{ x \in \mathbb{R}^n : Ax \leq b \right\}$$

is closed since it is the intersection of closed half-spaces. The maximum is therefore attained if $P(A, b)$ is non-empty and the set $\{cx : Ax \leq b\}$ is bounded above.

A *feasible solution* of the linear program (4.1) is a point $x \in P(A, b)$. A feasible solution at which the objective function c attains the maximum is called an *optimal solution*. An optimal solution is in general not unique.

M. Joswig, T. Theobald, *Polyhedral and Algebraic Methods in Computational Geometry*, 47
Universitext, DOI 10.1007/978-1-4471-4817-3_4,
© Springer-Verlag London 2013

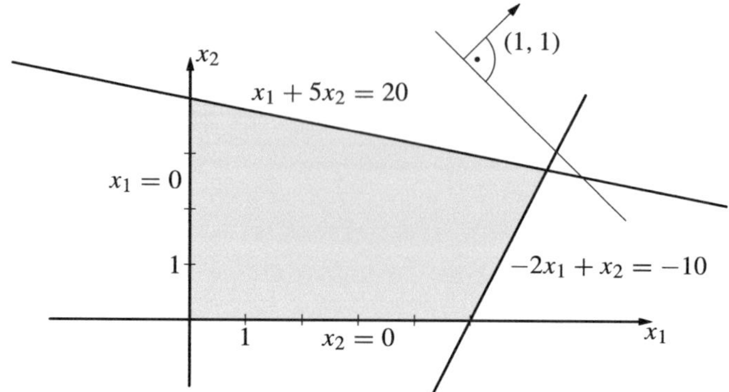

Fig. 4.1   A linear program in dimension 2

We formulate the linear programming problem stated at the beginning of this chapter as follows.

Input: A matrix $A \in \mathbb{R}^{m \times n}$ and vectors $b \in \mathbb{R}^m$, $c \in (\mathbb{R}^n)^*$.
Output: Either a vector $x \in \mathbb{R}^n$, such that $Ax \leq b$ and cx is maximized, or a statement declaring if $P(A, b)$ is empty or c is not bounded above on $P(A, b)$.

In the following we always assume that $c \neq 0$. Otherwise, the linear programming problem reduces to a pure feasibility problem, i.e., finding an arbitrary feasible point. We will return to the linear feasibility problem at the end of this chapter.

Exercise 4.1 Show that the set of all optimal solutions of $\max\{cx : Ax \leq b\}$ is a face of $P(A, b)$. Which conditions guarantee that the optimal solutions form a proper face?

Example 4.2 Maximizing the linear objective function $(1, 1)x$ for $x \in \mathbb{R}^2$ such that $x_1 + 5x_2 \leq 20$, $-2x_1 + x_2 \leq -10$, $x \geq 0$ can be written in normal form (4.1) as

$$\max(1, 1)x$$
$$\begin{pmatrix} 1 & 5 \\ -2 & 1 \\ -1 & 0 \\ 0 & -1 \end{pmatrix} x \leq \begin{pmatrix} 20 \\ -10 \\ 0 \\ 0 \end{pmatrix}.$$

Figure 4.1 depicts the feasible region of this linear program. The maximal value of the objective function is $100/11$ and is attained at the point $(70/11, 30/11)$. The objective function c is constant on each line satisfying the equation $(1, 1)x = \alpha$ for $\alpha \in \mathbb{R}$. Here, the optimal solution is unique. It is the intersection of P with the line from the set of lines $(1, 1)x = \alpha$ which has maximal α and still intersects P.

Before we study the solution of linear programs in general, we look at a particular application that will be very useful to us later.

Example 4.3 Let

$$H_i^+ = \left[h_0^{(i)} : \cdots : h_n^{(i)} \right]^+ \quad \text{for } 1 \leq i \leq m$$

be affine half-spaces in $\mathbb{R}^n$. We search for an interior point of the polyhedron $P = \bigcap_{i=1}^m H_i^+$, or alternatively verification that $\dim P < n$, in which case the interior is empty. To do this, we study the linear program

$$h_0^{(1)} + h_1^{(1)} x_1 + \cdots + h_n^{(1)} x_n \geq \epsilon,$$
$$\vdots \tag{4.2}$$
$$h_0^{(m)} + h_1^{(m)} x_1 + \cdots + h_n^{(m)} x_n \geq \epsilon.$$

Clearly, we have that $x = (x_1, \ldots, x_n)^T$ is an interior point of P if and only if there exists an $\epsilon > 0$ such that (4.2) holds. Therefore, the question of the existence of an interior point of the polyhedron P can be answered by maximizing ϵ under the linear conditions (4.2). If the maximal ϵ is positive, we know that the corresponding point x is an interior point and that $\dim P = n$. If $\epsilon = 0$ we have that $P \neq \emptyset$ and $\dim P < n$. If $\epsilon < 0$ then $P = \emptyset$.

The unbounded case is also possible, i.e., there may exist an arbitrarily large $\epsilon > 0$ that satisfies the conditions. This can easily be avoided by introducing the artificial constraint $\epsilon \leq 1$.

Exercise 4.4 Let $P = \bigcap_{i=1}^m H_i^+$ be given in $\mathcal{H}$-representation.

(a) Construct a linear program which enables you to describe an affine hyperplane that contains P, or which enables you to determine that such a hyperplane does not exist.
(b) Describe a method to compute the dimension of P.

Exercise 4.5 Let $P = \bigcap_{i=1}^m H_i^+$ be given in $\mathcal{H}$-representation. Describe a method to compute the lineality space of P.

4.2 Duality

Using duality theory it is possible to characterize when a given point is an optimal point of a linear program. To do this, we will first examine the geometry in greater detail.

The feasible region of the linear program $\max\{cx : Ax \leq b\}$ is the polyhedron $P := P(A, b) \subseteq \mathbb{R}^n$. By Exercise 4.1 we know that any optimal solution must be on the boundary of P. Given an arbitrary point v on the boundary of the polytope P, let $(A'(v) \mid b'(v))$ be the submatrix of $(A \mid b)$ consisting of those rows which correspond to inequalities that are satisfied by v as equalities; these are called *active* in v. Since v lies on the boundary, there is at least one active inequality in v. Note that $\partial P(A, b) = P(A, b)$ holds if $\dim P(A, b) < n$.

The inactive inequalities are summarized in the matrix $(A''(v) \mid b''(v))$. Up to reordering of rows we may assume that

$$(A \mid b) = \begin{pmatrix} A'(v) \mid b'(v) \\ A''(v) \mid b''(v) \end{pmatrix}$$

and

$$A'(v)v = b'(v), \tag{4.3}$$

$$A''(v)v < b''(v), \tag{4.4}$$

for all $v \in \partial P$. We will see in the following that for each optimal solution v of the LP $\max\{cx : Ax \le b\}$ the point v is also an optimal solution of the LP

$$\max\{cx : A'(v)x \le b'(v)\}.$$

The cone

$$N(v) := \mathrm{pos}\{a_1(v), \ldots, a_k(v)\} \subseteq (\mathbb{R}^n)^*$$

generated by the rows $a_1(v), \ldots, a_k(v)$ of the matrix $A'(v)$ is called the *outer normal cone* in v.

Exercise 4.6 Show that for $v \in \partial P$ the following statements are equivalent:

(a) The point v is a vertex of P.
(b) The matrix $A'(v)$ of active conditions has full rank n.

If P is full-dimensional and v is a vertex of P, then the cone $N(v)$ is pointed.

Example 4.7 The left hand picture in Fig. 4.2 shows the cone $N(v)$ of outer normals in a vertex $v = 0$ in a triangle given as the intersection of three half-spaces. The boundary rays of the cone are perpendicular to the two lines incident with v. If v is a non-zero point, we obtain the cone of outer normals as a translation of the depicted cone.

The right hand picture shows the intersection of four half-spaces such that the vertex v satisfies three of the four inequalities as an equality. The cone $N(v)$ is the same as before.

Lemma 4.8 *Let v be a boundary point of the polyhedron $P = \{x \in \mathbb{R}^n : Ax \le b\}$. Then*

$$N(v) = \{u \in (\mathbb{R}^n)^* : ux \le 0 \text{ for all } x \in P(A'(v), 0)\}.$$

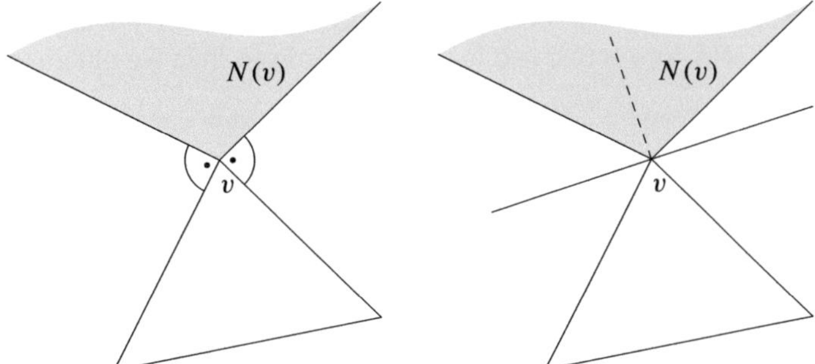

Fig. 4.2 The cone $N(v)$ of outer normals where the origin of the dual space $(\mathbb{R}^2)^*$ has been moved to the point v of the original space $\mathbb{R}^2$

Proof We denote the rows of $A'(v)$ by $a_1, \ldots, a_k$.

"$\subseteq$": Let $u \in N(v)$. Then there exist $\lambda_1, \ldots, \lambda_k \geq 0$ with $u = \sum_{i=1}^{k} \lambda_i a_i$. We therefore have for every point $x \in P(A'(v), 0) = \{x \in \mathbb{R}^n : A'(v)x \leq 0\}$

$$ux = \sum_{i=1}^{k} \lambda_i a_i x \leq 0.$$

"$\supseteq$": Let $u \notin N(v)$. By the separation theorem there exists a vector $w \in \mathbb{R}^n$ with $uw > 0$, but $zw \leq 0$ for all $z \in N(v)$. In particular, we infer that $a_i w \leq 0$ for all $1 \leq i \leq k$, and thus $A'(v)w \leq 0$. This implies $w \in P(A'(v), 0)$. Since $uw > 0$, we have

$$u \notin \left\{ y \in \left(\mathbb{R}^n\right)^* : yx \leq 0 \text{ for all } x \in P\big(A'(v), 0\big) \right\}. \qquad \square$$

Corollary 4.9 *Let* $P = \{x \in \mathbb{R}^n : Ax \leq b\}$, $c \in (\mathbb{R}^n)^* \setminus \{0\}$ *and* v *be a boundary point of* P. *The point* v *is an optimal solution of the LP* $\max\{cx : x \in P\}$ *if and only if* c *is contained in the cone* $N(v)$ *of outer normals in* v.

Proof The point $v \in \partial P$ is optimal for the LP $\max\{cx : x \in P\}$ if and only if all $x \in P$ satisfy $cx \leq cv$, i.e., $c(x - v) \leq 0$. Thus, the previous lemma implies the statement. $\qquad \square$

In other words, a point $v \in P$ is an optimal solution of the LP $\max\{cx : Ax \leq b\}$ if and only if the system of inequalities in the variables $y = (y_1, \ldots, y_m)$

$$yA = c,$$
$$y \geq 0 \tag{4.5}$$

has a solution in $(\mathbb{R}^m)^*$ for which only those components of y corresponding to active conditions for v are non-zero. If we assume that v is a vertex of the polyhedron P, then by Exercise 4.6 the submatrix $A'(v)$ is invertible. Furthermore,

$A'(v)v = b'(v)$. If $y = (y', y'')$ denotes the decomposition of y with respect to the decomposition of A into active and inactive conditions, then we obtain the solution to (4.5) via

$$y' = cA'(v)^{-1},$$
$$y'' = 0.$$
$$(4.6)$$

Definition 4.10 Given the linear program $\max\{cx : Ax \leq b\}$, then

$$\begin{aligned} \min \quad & yb \\ & yA = c, \\ & y \geq 0 \end{aligned} \qquad (4.7)$$

is the *dual program*. The original problem $\max\{cx : Ax \leq b\}$ is also referred to as the *primal program*. A feasible solution to the primal LP is said to be *primal feasible* and the term *dual feasible* is analogously defined.

Theorem 4.11 (Weak Duality Theorem) *If x is a primal feasible solution and y is a dual feasible solution, then $cx \leq yb$.*

Proof For primal and dual feasible solutions x and y we have

$$cx = (yA)x = y(Ax) \leq yb. \qquad \square$$

A pair (x, y) of feasible solutions of the dual LPs

$$\max\{cx : Ax \leq b\} \quad \text{and} \quad \min\{yb : yA = c, \ y \geq 0\} \qquad (4.8)$$

is called a *primal-dual pair* if the *complementary slackness* condition

$$y(b - Ax) = 0$$

is satisfied.

Exercise 4.12 Let (x, y) be a primal-dual pair. Show that x is an optimal point of the primal LP and y is an optimal point of the dual LP.

Theorem 4.13 (Strong Duality Theorem) *For a pair of dual linear programming problems*

$$\max\{cx : Ax \leq b\} \quad \text{and} \quad \min\{yb : yA = c, \ y \geq 0\}$$

exactly one of the following conditions holds:

(a) *Both problems are feasible and the optimal values are the same.*
(b) *One of the problems is infeasible and the other one is unbounded.*
(c) *Both problems are infeasible.*

Proof We assume that the primal problem is feasible and bounded. Let $v \in \mathbb{R}^n$ be an optimal point of the primal problem. Then by Corollary 4.9 there exists a vector $y = (y_1, \ldots, y_m)$ such that

$$yA = c,$$

$$y \geq 0,$$

and only components corresponding to the active conditions of v can be non-zero. Thus, we have

$$cv = (yA)v = y(Av) = yb.$$

Now the weak duality theorem implies that the dual problem is bounded and that the optimal values are equal.

The cases in which one of the problems is infeasible or unbounded are left as an exercise for the reader. $\qquad\qquad\square$

4.3 The Simplex Algorithm

The most famous algorithm for linear programming is the *simplex algorithm* (Dantzig, 1947). Let the linear program be given as before in the form $\max\{cx : x \in P\}$ with $P = P(A, b) = \{x \in \mathbb{R}^n : Ax \leq b\}$. We first assume that the polyhedron is full-dimensional and pointed, and that one vertex of P (the "start vertex") is already known. In particular, P is non-empty. We will later show that these assumptions are justified.

The simplex algorithm relies on a simple geometric idea. First, we check if the current vertex v of P is an optimal point. For this, the duality theory from the previous section will be very useful. If v is not an optimal point, we will compute those edges starting at v with respect to which the objective function increases. In this way, we either find a "better" vertex of P or an unbounded edge (see Fig. 4.3). This method ends after finitely many steps, since P has only finitely many vertices.

Let v be a vertex of P. The equivalent conditions from Exercise 4.6 imply that rank $A'(v) = n$.

Definition 4.14 A linearly independent subset of rows of $A'(v)$ which spans the row-space of $A'(v)$ is called a *basis* for v with respect to A.

If P is a simple polytope and the rows of the matrix A consist of the unique (up to scaling) outer facet normals of P, then the basis is uniquely determined for every vertex v. Every basis of v defines a pointed cone with apex v that contains the polyhedron P. This cone is projectively equivalent to a simplex, which can be viewed as a local approximation of the polyhedron P in v. This is the reason why the method is called the simplex algorithm.

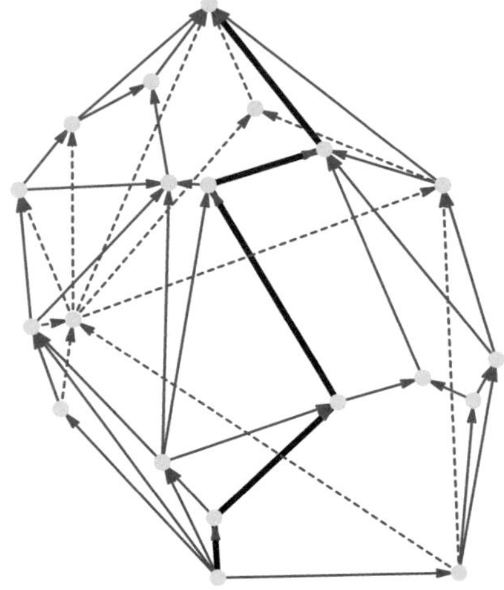

Fig. 4.3 A possible path of the simplex algorithm in $\mathbb{R}^3$. The arrows on the edges indicate the direction induced by the objective function

Example 4.15 Let P be the polytope

$$P = \left\{x \in \mathbb{R}^3 : 0 \le x_i \le 1 \text{ for } 1 \le i \le 3, \ x_1 + 2x_2 + x_3 \le 3\right\}$$

contained in the unit cube $[0, 1]^3$. The inequalities can be expressed in matrix form as

$$\begin{pmatrix} -1 & 0 & 0 \\ 0 & -1 & 0 \\ 0 & 0 & -1 \\ 1 & 0 & 0 \\ 0 & 1 & 0 \\ 0 & 0 & 1 \\ 1 & 2 & 1 \end{pmatrix} x \le \begin{pmatrix} 0 \\ 0 \\ 0 \\ 1 \\ 1 \\ 1 \\ 3 \end{pmatrix} \tag{4.9}$$

(see Fig. 4.4). We write (4.9) as $Ax \le b$ for short. The vertices $v = (0, 0, 0)^T$ and $w = (1, 1, 0)^T$ satisfy

$$A'(v) = \begin{pmatrix} -1 & 0 & 0 \\ 0 & -1 & 0 \\ 0 & 0 & -1 \end{pmatrix} \quad \text{and} \quad A'(w) = \begin{pmatrix} 0 & 0 & -1 \\ 1 & 0 & 0 \\ 0 & 1 & 0 \\ 1 & 2 & 1 \end{pmatrix}.$$

The rows of $A'(v)$ define the unique basis of v with respect to A and each set of three rows of $A'(w)$ is a basis of w.

We will now describe the main step of the simplex algorithm. In the following, let $v \in \mathbb{R}^n$ be a vertex of P and I be the set of row indices of a basis of v. We denote by A_I the submatrix of A induced by I and use the corresponding notation for the vector b. In particular, in this notation we have $b_{\{i\}} = b_i$. By our assumption we have

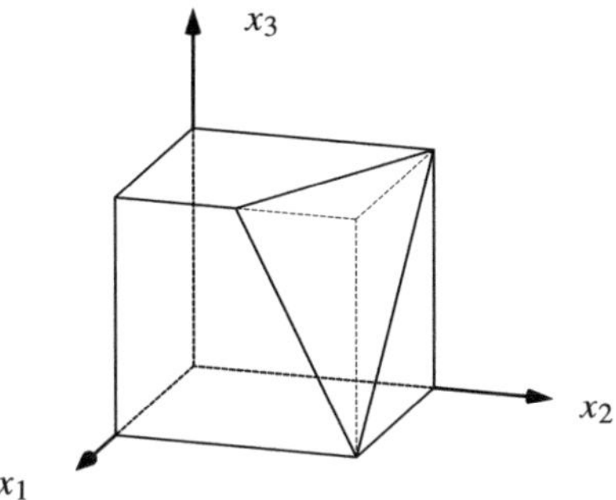

Fig. 4.4 The polytope P contained in the cube $[0, 1]^3$

that A_I is regular and that $A_I v = b_I$. Let a_i denote the i-th row of the matrix A, i.e., $a_i = A_{\{i\}}$. Every edge of the approximating cone $K_I = \{x \in \mathbb{R}^n : A_I x \leq b_I\}$ is the intersection of exactly $n - 1$ facets of K_I, i.e., every row $i \in I$ defines a non-pointed cone $K_{I \setminus \{i\}}$ whose one dimensional lineality space contains an edge of K_I.

Lemma 4.16 *The set* $L = \{x \in \mathbb{R}^n : A_{I \setminus \{i\}} x = b_{I \setminus \{i\}}\}$ *is an affine line in* $\mathbb{R}^n$ *that contains* v. *Furthermore, the column of* $-(A_I)^{-1}$ *with index* i *is a directional vector of* L.

Proof The regularity of A_I implies that L is a line and, clearly, the point v lies on L. Let s be the column of $-(A_I)^{-1}$ with index i. Then we have

$$A_{I \setminus \{i\}} s = 0 \quad \text{and} \quad a_i s = -1. \tag{4.10}$$

Thus, s is a non-zero vector such that $v + s \in L$. $\square$

Now let s be the column of $-(A_I)^{-1}$ with index i. Starting from the vertex v, we could search for better solutions in the direction of s. The usefulness of this method depends on whether the objective function increases in the direction of s, i.e., if $cs > 0$. This can be determined via the dual program.

Lemma 4.17 *Let* $y \in (\mathbb{R}^m)^*$ *with* $yA = c$ *and* $y_j = 0$ *for all* $j \notin I$. *Then* $cs > 0$ *if and only if* $y_i < 0$.

Proof Let y denote such a dual feasible solution. Then the definition of the dual program and (4.10) yield that

$$cs = yAs = y_I A_I s_I = -y_i.$$ $\square$

Now the idea is to walk from the vertex v in the direction of s on a suitable edge of K_I as long as we do not violate the feasibility conditions. When doing this two cases can occur.

Lemma 4.18

(a) *If $As \leq 0$ then $v + \lambda s$ is feasible for all $\lambda \geq 0$.*
(b) *Otherwise, for*

$$\lambda_s := \min_j \left\{ \frac{b_j - a_j v}{a_j s} : a_j s > 0 \right\} \tag{4.11}$$

the point $v + \lambda_s s$ is feasible and λ_s is maximal with this property.

Proof To show this we first examine an arbitrary row index j and the corresponding condition $a_j v \leq b_j$.

Claim We have

$$\max\left\{ \lambda \geq 0 : v + \lambda s \in \{x \in \mathbb{R}^n : a_j x \leq b_j\} \right\} = \begin{cases} \frac{b_j - a_j v}{a_j s} & \text{if } a_j s > 0, \\ \infty & \text{if } a_j s \leq 0. \end{cases}$$

Since v is a feasible point, $a_j v \leq b_j$. If $a_j s \leq 0$, then for all $\lambda \geq 0$ the inequality $a_j(v + \lambda s) \leq a_j v \leq b_j$ holds. However, $a_j(v + \lambda s) \leq b_j$ holds if and only if $\lambda \leq (b_j - a_j v)/(a_j s)$.

To determine when $v + \lambda s$ violates the feasibility conditions of P, we test all inequalities simultaneously and in this way obtain the statement of the lemma. $\square$

Since the λ_s that was chosen in the case $As \not\leq 0$ in Lemma 4.18 was maximal, we infer that for $\lambda_s > 0$ one inequality that previously was not active becomes active at the point $v + \lambda_s s$.

Lemma 4.19 *Let j be a row index of the matrix A with $\lambda_s = (b_j - a_j v)/(a_j s)$. Then $v' := v + \lambda_s s$ is a vertex of P and $(I \setminus \{i\}) \cup \{j\}$ is the index set of a basis for v'.*

Proof Let $I' = (I \setminus \{i\}) \cup \{j\}$. We have to show that $A_{I'}$ is regular and that $A_{I'} v' = b_{I'}$.

Since $A_{I \setminus \{i\}} s = 0$ and $a_j s > 0$, a_j does not lie in the row-space of the $(n-1)$-row matrix $A_{I \setminus \{i\}}$. Thus $A_{I'}$ is regular.

From $A_{I \setminus \{i\}} s = 0$ and $\lambda_s = (b_j - a_j v)/(a_j s)$ follows

$$A_{I \setminus \{i\}}(v + \lambda_s s) = A_{I \setminus \{i\}} v = b_{I \setminus \{i\}}$$

and

$$a_j(v + \lambda_s s) = a_j v + a_j s \frac{b_j - a_j v}{a_j s} = b_j.$$

Hence we have $A_{I'} v' = b_{I'}$. $\square$

Example 4.20 We again examine the polytope from Example 4.15 and the objective function vector $(0, 0, 1)$. The set $I = \{3, 4, 7\}$ is the index set of a basis of the vertex $(1, 1, 0)^T$. For $i = 3$ we have for the line L from Lemma 4.16 that

$$L = \left\{ x \in \mathbb{R}^3 : x_1 = 1, \ x_1 + 2x_2 + x_3 = 3 \right\}.$$

This implies

$$-(A_I)^{-1} = -(A_{\{3,4,7\}})^{-1} = \begin{pmatrix} 0 & -1 & 0 \\ -\frac{1}{2} & \frac{1}{2} & -\frac{1}{2} \\ 1 & 0 & 0 \end{pmatrix},$$

so that the column corresponding to $i = 3$ is $s = (0, -1/2, 1)^T$. Furthermore, the equation $\lambda_s = 1$ holds, and the minimum in (4.11) is attained for the index $j = 6$. The index set of the new basis is therefore $I' = \{4, 6, 7\}$, and the new vertex is $v' = (1, 1/2, 1)^T$.

The λ_s that we defined in (4.11) could be zero. Only in the case $\lambda_s > 0$ will we find a vertex for which, when we start at v and travel in the direction of s, the objective function is improved. A vertex can have several bases, which can lead to $\lambda_s = 0$.

Example 4.21 Consider the previously discussed example. When we start from the basis of the vertex $(1, 1, 0)^T$ with index set $\{3, 4, 5\}$ and choose $i = 3$, we obtain $s = (0, 0, 1)^T$. The objective function c increases in direction s, since $cs = (0, 0, 1) \cdot (0, 0, 1)^T = 1$. Due to the inequality $x_1 + 2x_2 + x_3 \leq 1$ that corresponds to the last row of A, we have that $\lambda_s = 0$, which implies $v' = v$. Hence, we did not make a real gain. There has only been a so-called *change of basis*.

Algorithm 4.1 is a precursor of the simplex algorithm. *If it terminates*, it finds either an optimal vertex, or shows that the LP is unbounded.

The existence of a basis in Step 1 follows, according to Exercise 4.6, from the fact that in a vertex at least n (linear independent) inequalities have to be satisfied as equalities. First, all active inequalities are determined. Then we can compute a basis of the dual space $(\mathbb{R}^n)^*$ using Gaussian elimination. Determining the vector y in Step 2 can be performed using (4.6).

Theorem 4.22 *If Algorithm* 4.1 *terminates, we have the following: If it yields v and y in Step 4, then these vectors are optimal solutions to the dual LPs, and we have $cv = yb$. If the algorithm yields s in Step 8, then $cs > 0$, and the LP is unbounded.*

The case of an infeasible problem does not occur in this theorem since we assumed that we are given a start vertex.

Proof In Step 2 of the algorithm we determine a feasible solution of the dual feasibility space via (4.6). After that, we choose in Step 6 a search direction s according

---

Algorithm 4.1: A precursor of the simplex algorithm

Input: A matrix $A \in \mathbb{R}^{m \times n}$ and vectors $b \in \mathbb{R}^m$, $c \in (\mathbb{R}^n)^*$; a vertex v of
$\qquad P = \{x \in \mathbb{R}^n : Ax \le b\}$.
Output: An optimal vertex v of P (and dual vector y), or a vector $s \in \mathbb{R}^n$ with
$\qquad As \le 0$ and $cs > 0$ (i.e., the LP is unbounded).

1   $I \leftarrow$ index set of a basis for v
2   Determine a $y \in (\mathbb{R}^m)^*$ with $yA = c$ and $y_i = 0$ for all $i \notin I$.
3   **if** $y \ge 0$ **then**
4      $\lfloor$ **return** (v, y)
5   $i \leftarrow$ an index with $y_i < 0$
6   $s \leftarrow$ column of $-(A_I)^{-1}$ with index i, such that $A_{I \setminus \{i\}} s = 0$ and $a_i s = -1$
7   **if** $As \le 0$ **then**
8      $\lfloor$ **return** s
9   $\lambda_s \leftarrow \min_j \{ \frac{b_j - a_j v}{a_j s} : a_j s > 0 \}$; $j \leftarrow$ a row index that attains this minimum
10 $I \leftarrow (I \setminus \{i\}) \cup \{j\}$; $v \leftarrow v + \lambda_s s$
11 **goto** Step 2

---

to Lemma 4.16. By Lemma 4.17 we have $cs > 0$, so the value of the objective
function is improved in this direction. We compute the maximal step-size λ_s using
Lemma 4.18 in Step 9 and after that we obtain, by Lemma 4.19, the new basis in
Step 10.

If the algorithm yields v and y in Step 4, then v and y form a primal-dual pair:
We have $cv = (yA)v = y(Av) = yb$, since the components of y that lie outside the
index set I are zero. The weak duality theorem implies that v and y are optimal.

If the algorithm terminates in Step 8, then the LP is unbounded, since in this case
we have $cs > 0$ and thus $v + \lambda s \in P$ for all $\lambda \ge 0$. $\qquad\qquad\square$

If we choose at each step an arbitrary i with $y_i < 0$ and an arbitrary j, then
it may happen that, in the case of vertices with a non-unique basis, the algorithm
becomes trapped in a cyclic repetition, and therefore does not terminate. With a
suitable choice of the indices i and j, we can be certain that a non-optimal vertex is
left after finitely many steps. The most famous rule of this kind (the "pivot rule") is
the *rule of Bland*. In this rule we choose, in the case of multiple choices, the indices
i and j in Steps 5 and 9 to be minimal. Algorithm 4.2 describes the simplex method
with Bland's pivot rule.

Theorem 4.23 *The simplex algorithm terminates after at most $\binom{m}{n}$ iterations and
the conclusions of Theorem 4.22 hold.*

The proof uses only elementary facts but is somewhat tricky.

Algorithm 4.2: Modifications for the precursor of the simplex Algorithm 4.1 using Bland's pivot rule

---

5 $i \leftarrow$ minimal index such that $y_i < 0$

9 $\lambda_s \leftarrow \min\limits_{j}\left\{\frac{b_j - a_j v}{a_j s} : a_j s > 0\right\}$; $j \leftarrow$ smallest row index that attains the

maximum

Proof Let $I^{(k)}$ and $v^{(k)}$ be the index set and the vertex v in the k-th iteration of the simplex algorithm respectively. We denote the corresponding instances of the other variables analogously.

If the algorithm does not terminate after $\binom{m}{n}$ iterations, then there exists $k < l$ with $I^{(k)} = I^{(l)}$ and therefore $v^{(k)} = v^{(l)}$. Since the method always searches in an increasing direction with respect to the objective function, cv does not decrease in any iteration, and for a positive step-size λ_s in Step 9 it actually increases. Thus, in the iterations $k, k+1, \ldots, l-1$, we have that $\lambda_{s(k)} = \cdots = \lambda_{s(l-1)} = 0$, which implies $v^{(k)} = v^{(k+1)} = \cdots = v^{(l)}$. Let h be the maximal index which is taken out of a basis I in one of the iterations $k, \ldots, l-1$, and assume that this happens in iteration p. Since $I^{(k)} = I^{(l)}$, we know that the index h must have been added to I in an iteration $q \in \{k, \ldots, l-1\}$. Hence, we obtain in particular that $a_h s^{(q)} > 0$.

Since $c = y^{(p)} A$, we have $y^{(p)} A s^{(q)} = c s^{(q)} > 0$. Hence, there exists an $r \in \{1, \ldots, m\}$ such that $y_r^{(p)} a_r s^{(q)} > 0$, and thus in particular $y_r^{(p)} \neq 0$. By Step 2 of the algorithm this implies $r \in I^{(p)}$, because all components of $y^{(p)}$ outside of $I^{(p)}$ vanish.

Now consider the cases $r > h$ and $r \leq h$. In the case $r > h$, the index r will never be taken out of the basis, and thus, in iteration q, we have $a_r s^{(q)} = 0$ due to Step 6 of the algorithm. This contradicts our definition of r.

In the case $r \leq h$, Bland's rule implies that in Step 5 of iteration p, we have $y_r^{(p)} < 0$ if and only if r coincides with h. Furthermore, Bland's rule implies that in Step 9 of iteration q, we have $a_r s^{(q)} > 0$ if and only if r coincides with h. Both in the case $r = h$ and in the case $r < h$ we therefore have $y_r^{(p)} a_r s^{(q)} \leq 0$, which contradicts $y_r^{(p)} a_r s^{(q)} > 0$.

We showed that no basis is attained multiple times and that $\binom{m}{n}$ is an upper bound for the number of possible bases. $\square$

The identification of bases with the sets of row indices of the matrix A induces an order on the bases that are visited. The bases which are visited during a run of the simplex algorithm with Bland's pivot rule are strictly increasing with respect to this order.

There exist examples with n variables and $2n$ linear conditions (the "Klee–Minty cube") for which the simplex algorithm (with Bland's pivot rule) needs exponentially many iterations in n. This shows that the run-time of the simplex algorithm with Bland's pivot rule is not polynomially bounded in the dimension. There are many pivot rules which describe how to choose i and j; however, whether there exists a pivot rule that leads to a polynomial time algorithm is a very important open problem.

4.4 Determining a Start Vertex

Thus far we have always assumed that our linear program is feasible and that we already know a feasible vertex. We want to clarify the general case in the following.

We examine the linear program with non-negativity conditions:

$$\max\{cx : Ax \le b, \ x \ge 0\}. \tag{4.12}$$

This is not a significant restriction, since every linear program of the form $\max\{cx : Ax \le b\}$, as in (4.1), can be transformed to the form (4.12) in the following way. Every vector $x \in \mathbb{R}^n$ has a (in general not unique) representation of the form $x = x^+ - x^-$ with $x^+, x^- \in \mathbb{R}^n_{\ge 0}$. We replace x by $x^+ - x^-$ and write the LP in the form

$$\begin{aligned} \max \quad & (c, -c) \begin{pmatrix} x^+ \\ x^- \end{pmatrix} \\ & (A, -A) \begin{pmatrix} x^+ \\ x^- \end{pmatrix} \le b, \\ & x^+, x^- \ge 0. \end{aligned} \tag{4.13}$$

This linear program in $2n$ variables is feasible if and only if the original program is feasible. In the feasible case, either both programs have the same optimal value or they are both unbounded. Note, however, that (4.13) has an unbounded feasible region whenever the feasible region of the original LP is non-empty.

We can therefore assume in the following that we are given an LP in the form (4.12), where A is an $m \times n$-matrix, $b \in \mathbb{R}^m$ and $c \in (\mathbb{R}^n)^*$. With the notation $I = \{i \in \{1, \ldots, m\} : b_i \ge 0\}$ and $J = \{j \in \{1, \ldots, m\} : b_j < 0\}$ we study the auxiliary problem

$$\begin{aligned} \min \quad & (\mathbf{1}A_J)x + \mathbf{1}y \\ & A_I x \le b_I, \\ & A_J x + y \ge b_J, \\ & x, y \ge 0, \end{aligned} \tag{4.14}$$

where $\mathbf{1}$ denotes the all-ones vector. Setting $k := |J|$, the linear program (4.14) has $n + k$ variables. Let $P' \subseteq \mathbb{R}^{n+k}$ be the feasible region of (4.14).

Proposition 4.24 *The origin is a vertex of P'. The minimal value μ of the auxiliary problem is finite and we have $\mu \ge \mathbf{1}b_J$. If $\mu > \mathbf{1}b_J$, then (4.12) is infeasible. If $\mu = \mathbf{1}b_J$, then for each optimal vertex $\begin{pmatrix} x^* \\ y^* \end{pmatrix}$ of the auxiliary problem the point x^* is a vertex of the feasible region of (4.12).*

Proof The origin is contained in P' by definition. Since the non-negativity conditions $x_i \geq 0$ and $y_i \geq 0$ are active in 0, we know that the origin is a vertex of P'.

The objective function of the auxiliary problem is bounded below by $\mathbf{1}b_J$, i.e., $\mu \geq \mathbf{1}b_J$. Thus, for every feasible solution x of (4.12) the choice $y := b_J - A_J x$ yields an optimal solution $\binom{x}{y}$ of (4.14). This shows that the LP (4.12) is infeasible for $\mu > \mathbf{1}b_J$.

Now let $\mu = \mathbf{1}b_J$ and $\binom{x^*}{y^*}$ be an optimal vertex of (4.14). Since the objective function minimizes the sum of the entries of the vector $A_J x + y$, we note $A_J x^* + y^* = b_J$, and x^* is therefore feasible for (4.12). By Exercise 4.6, we can pick a set S of $n + k$ independent inequalities which are active in x^*.

Denote by S_I the set of inequalities among $A_I x \leq b_I$ or among $x \geq 0$ that are contained in S. Denote by S_J the set of inequalities among $A_J x \leq b_J$ such that the corresponding inequalities of $A_J x + y \leq b_J$ and of $y \geq 0$ are both contained in S. Note that any inequality from S_I and any inequality from S_J provides an active inequality in x^* for the original problem (4.12). By the independence of the inequalities in S_I and S_J it therefore suffices to show $|S_I \cup S_J| \geq n$. In order to see this, observe that the definition of S_J implies $|S_I \cup S_J| \geq |S| - |S_J| \geq |S| - k = n$. Thus we have found n independent active inequalities in x^*, which shows that x^* is a vertex of (4.12). $\qquad\square$

The most important consequence of this proposition is that the simplex algorithm can be used for the auxiliary problem, with start vertex 0, to decide if the original problem is feasible. If it is feasible, this method yields a start vertex for the simplex algorithm for the original LP (4.12).

Exercise 4.25 Apply the method for computing a start vertex to the linear program from Example 4.2.

4.5 Inspection Using `polymake`

`polymake` can also be used to solve linear programs (via interfaces to the libraries `cddlib` [43] and `lrslib` [6]). As an example we study here the auxiliary problem for a 2-dimensional LP in the form (4.12).

Let

$$A = \begin{pmatrix} 1 & 1 \\ -2 & -1 \end{pmatrix}, \qquad b = \begin{pmatrix} 2 \\ -1 \end{pmatrix} \quad \text{and} \quad c = (1, 0).$$

We want to solve the LP

$$\max\{cx : Ax \leq b, x \geq 0\}. \tag{4.15}$$

The auxiliary problem (4.14) has the unknowns x_1, x_2 and y, since the vector b has exactly one active component. Since an inequality of the form $u_0 + \sum_{i=1}^{n} u_i x_i \geq 0$

is represented in `polymake` as the homogeneous vector $(u_0, \ldots, u_n)$, we use the following input:

```
polytope > $aux_constraints
    = new Matrix([[2,-1,-1,0],[1,-2,-1,1],
                  [0,1,0,0],[0,0,1,0],[0,0,0,1]]);
```

Here, the first two inequalities in the matrix `$aux_constraints` correspond to the first two rows of A. The last three inequalities are the non-negativity conditions for x_1, x_2, y. The linear objective function is encoded into an object of type `LinearProgram` which is then attached to the polyhedron defined by the constraint matrix.

```
polytope > $aux_obj
    = new LinearProgram(LINEAR_OBJECTIVE=>[0,-2,-1,1]);
polytope > $aux = new Polytope(INEQUALITIES=>$aux_constraints,
                              LP=>$aux_obj);
```

Everything that we want to know about this linear program is now a property of the `$aux` object. Notice that, despite the name, objects of type `Polytope` may, in fact, be unbounded polyhedra.

```
polytope > print $aux->LP->MINIMAL_VALUE;
-1
polytope > print $aux->LP->MINIMAL_VERTEX;
1 0 2 1
```

The minimal value is $-1 = \mathbf{1}b_J$, which implies that the point $\binom{0}{2}$ is a feasible vertex of the original problem, which turns out to be not optimal. Actually, $\binom{2}{0}$ is the unique optimal solution of (4.15). If you are only interested in the solution of the LP (4.15), you do not need to explicitly construct the auxiliary problem.

```
polytope > $A = new Matrix([[1,1],[-2,-1]]);
polytope > $b = new Vector([2,-1]);
polytope > $c = new Vector([0,1,0]);
polytope > $non_negative = new Matrix([[0,1,0],[0,0,1]]);
polytope > $my_LP
   = new Polytope(INEQUALITIES=>($b|-$A)/$non_negative,
                  LP=>new LinearProgram(LINEAR_OBJECTIVE=>$c));
polytope > print $my_LP->LP->MAXIMAL_VALUE;
2
polytope > print $my_LP->LP->MAXIMAL_VERTEX;
1 2 0
```

To further analyze the geometry we can also check all of the vertices and list those which are maximal with respect to the given objective function.

```
polytope > print rows_numbered($my_LP->VERTICES);
0:1 0 1
1:1 1/2 0
2:1 0 2
3:1 2 0
```

```
polytope > print $my_LP->LP->MAXIMAL_FACE;
{3}
```

Here, the vertex with number 3 is the only optimal solution; to see to which point this actually refers, above we listed the vertices in numerical order.

4.6 Exercises

Exercise 4.26 (Farkas' lemma) Let A be an $m \times n$-matrix and $b \in \mathbb{R}^m$. Then *either* the system of inequalities

$$Ax = b, \quad x \geq 0 \quad (x \in \mathbb{R}^n)$$

or the system of inequalities

$$A^T z \geq 0, \qquad b^T z < 0 \quad (z \in \mathbb{R}^m)$$

has a solution.

[*Hint*: If the first system has no solution, then the set $\{y \in \mathbb{R}^m : Ax = y$ for an $x \geq 0\}$ can be strictly separated from the vector b.]

Exercise 4.27 Construct different dual pairs of linear programming problems

$$\max\{cx : Ax \leq b\} \quad \text{and} \quad \min\{yb : yA = c, \, y \geq 0\}$$

with the following additional characteristics:

(a) the primal problem is unbounded and the dual problem is feasible;
(b) the primal problem is infeasible and the dual is unbounded;
(c) both problems are infeasible.

Exercise 4.28 Two $\mathcal{V}$-polytopes in $\mathbb{R}^n$,

$$P = \mathrm{conv}\{p^{(1)}, \ldots, p^{(m)}\} \quad \text{and} \quad Q = \mathrm{conv}\{q^{(1)}, \ldots, q^{(r)}\},$$

are given. Formulate a linear program to determine if there is a separating hyperplane for P and Q. If it exists describe such a hyperplane.

An important special case of the previous exercise is the case when $P = \mathrm{conv}\{p^{(1)}, \ldots, p^{(m)}\}$ is arbitrary and $Q = \{x\}$ is just a point. A separating hyperplane exists in this case if and only if x is a vertex of $\mathrm{conv}(P \cup \{x\})$. In this way we obtain an LP-based method to determine the vertices of a polytope in the $\mathcal{V}$-representation.

4.7 Remarks

As mentioned above, it is unknown if there exists a suitable pivot rule that makes the simplex algorithm a polynomial time algorithm. However, there exist numerically efficient variants of the basic algorithm introduced here that are well suited for solving linear programs in real world applications. Most implementations work with the representation $\max\{cx : Ax = b, \ x \geq 0\}$, instead of the normal form (4.1) that we chose here.

There exist polynomial-time algorithms for solving linear programs: the ellipsoid method (Khachiyan, 1979), which is not suitable for practical applications, as well as interior point methods (Karmarkar, 1984). Currently, the interior point method seems to be the simplex algorithm's strongest competitor. Current research activity in this area, as well as improving programming techniques, do not allow us to make a final judgment on which algorithm for linear programming is the best from the perspective of real world applications.

In contrast to this, the problem of determining an optimal integer point in a polytope is NP-hard. In Section 10.6 we will introduce an algorithmic method that solves certain integer linear programs.

Our presentation of linear programming is based on the books by Gritzmann [53] and Korte and Vygen [72]. Further material can be found in the standard texts by Chvátal [22], Schrijver [91] and Grötschel, Lovász and Schrijver [54].

Chapter 5
Computation of Convex Hulls

When referring to "computation of convex hulls" we understand this as the task of computing the $\mathcal{H}$-representation of the convex hull of a given finite point set $V \subseteq \mathbb{R}^n$. Depending on the desired application, one might also need to compute all faces, a description of the face lattice or other geometric information.

5.1 Preliminary Considerations

We begin with two simple results. First, Algorithm 5.1 immediately gives a trivial convex hull algorithm, which is, unfortunately, inefficient.

Theorem 3.9, together with the fact that the computed half-spaces define facets, shows that the algorithm is correct. The assumption that the affine hull is full-dimensional is not necessary. Without it, the algorithm can simply be applied to the affine hull of the input.

Algorithm 5.1: A trivial convex hull algorithm

Input: Finite point set $V \subseteq \mathbb{R}^n$ with $\dim \operatorname{aff} V = n$.
Output: Finite set of half-spaces $\{H_1^+, \ldots, H_m^+\}$ such that
$$\bigcap_{i=1}^m H_i^+ = \operatorname{conv} V.$$

1   $\mathcal{H} \leftarrow \emptyset$
2   **foreach** n-element subset $W \subseteq V$ with $\dim \operatorname{aff} W = n-1$ **do**
3     $H \leftarrow \operatorname{aff} W$
4   **if** $V \subseteq H^+$ **then**
5       $\mathcal{H} \leftarrow \mathcal{H} \cup \{H^+\}$
6   **else**
7   **if** $V \subseteq H^-$ **then**
8         $\mathcal{H} \leftarrow \mathcal{H} \cup \{H^-\}$

9   **return** $\mathcal{H}$

M. Joswig, T. Theobald, *Polyhedral and Algebraic Methods in Computational Geometry*,   65
Universitext, DOI 10.1007/978-1-4471-4817-3_5,
© Springer-Verlag London 2013

Secondly, the dual problem, i.e., computing a $\mathcal{V}$-representation of a polytope from its $\mathcal{H}$-representation is, by polarity, algorithmically equivalent to the convex hull problem:

Theorem 5.1 *The problem of computing the $\mathcal{V}$-representation of a polytope from its $\mathcal{H}$-representation can be reduced to the convex hull problem and vice versa.*

Proof Let $P = \bigcap_{i=1}^{m} H_i^{+}$ be given in the $\mathcal{H}$-representation. Via the linear programs from Example 4.3 and Exercise 4.4 we can compute the affine hull $A = \operatorname{aff} P$ and a point from the relative interior x in $P = P \cap A$. We can thus assume that P is full-dimensional. We can also assume that the origin is an interior point, since we can otherwise apply our computations to A and translate by $-x$.

Since $0 \in \operatorname{int} P$, there exist $h_k^{(i)}$ such that $H_i^{+} = [1 : h_1^{(i)} : \cdots : h_n^{(i)}]^{+}$. We now examine the polar polytope which has, according to Theorem 3.28, the $\mathcal{V}$-representation $P^{\circ} = \operatorname{conv}\{h^{(1)}, \ldots, h^{(m)}\}$. Using a convex hull algorithm we can obtain an $\mathcal{H}$-representation $P^{\circ} = \bigcap_{j=1}^{k}[1 : v_1^{(j)} : \cdots : v_n^{(j)}]^{+}$. Looking at the polar polytope of P° and using Theorem 3.29 we get

$$P = P^{\circ\circ} = \operatorname{conv}\{v^{(1)}, \ldots, v^{(k)}\}.$$

The reverse direction is similar. In fact, it is easier since it is not necessary to use the linear programming techniques used above.   □

In the dual representation of the convex hull problem it becomes clear that the problem can be viewed as a far-reaching generalization of the linear optimization problem: While linear optimization aims at computing *one* specific vertex of an $\mathcal{H}$-polytope (defined by a linear objective function), the dual convex hull algorithm computes *all* vertices of P.

Note that the existence of cyclic polytopes of dimension n with m vertices and $\Theta(m^{\lfloor n/2 \rfloor})$ facets implies that there cannot exist a convex hull algorithm which is polynomial in m and n, since every such algorithm has to write the (in this case) exponentially many facets as output. Theorem 5.1 and the existence of the dual polytopes to cyclic polytopes imply that the dual convex hull problem has exponential run-time in the worst case. Now the natural question is if the naive algorithm from the beginning of this chapter can be optimized at all. There are two answers to this: First, by carefully analyzing the geometry we can exclude many hyperplanes which Algorithm 5.1 considers to be candidates for facets. We demonstrate how to do this in the next section. Secondly, the problem has a different quality when we assume the dimension n to be fixed: In Section 5.3 we will study the case $n = 2$. We provide further remarks and suggested literature at the end of this chapter.

5.2 The Double Description Method

To emphasize the relationship between the linear programming methods from the previous chapter and the convex hull problem, we study the convex hull problem in its dual form. A basic approach is to order the affine hyperplanes which were given as input. Our goal is to take $\mathcal{V}$-representations of polytopes which are intersections of k hyperplanes to obtain $\mathcal{V}$-representations of polytopes which are intersections of $k + 1$ hyperplanes. Such methods are called *iterative*. While reading this section, it is useful to think about how the specific steps can be translated into primal form.

Let P be an $\mathcal{H}$-polytope whose $\mathcal{V}$-representation $P = \operatorname{conv} V$ is already known. We now study how the $\mathcal{V}$-representation must be altered when another half-space H^+ is added. Define $P' = P \cap H^+$. The hyperplane H partitions the point set V into three parts: Points on the hyperplane and points on either of its two sides.

Lemma 5.2 *Let V_0, V_+, V_- be the partition of the point set V defined by*

$$V_0 = V \cap H, \qquad V_+ = V \cap H^+ \setminus H, \qquad V_- = V \cap H^- \setminus H.$$

Then we have

$$P' = \operatorname{conv}\big((V_0 \cup V_+) \cup \{[v, w] \cap H : v \in V_+, w \in V_-\}\big).$$

Proof It is obvious that the points in $V_0 \cup V_+$ are contained in P'. Furthermore, if $v \in V_+$ and $w \in V_-$, then the segment $[v, w]$ intersects the hyperplane H in one point which proves that $P' \supseteq \operatorname{conv} V'$.

For the reverse inclusion it is sufficient to examine the case where V is the vertex set of P. To find the vertices of P' we have to determine which cases have a supporting hyperplane of P' that intersects the polytope in exactly one point v. This happens when either v is a vertex of P (and contained in H^+) or v is the intersection of an edge of P with H. The edges of P are segments between vertices of P. The segment $[v, w]$ intersects the hyperplane H only in the two cases we mentioned, which proves the statement. $\square$

Using Lemma 5.2 we can immediately provide a method to iteratively transform an outer description of a polytope $P \subseteq \mathbb{R}^n$ into an inner description. Without loss of generality we again assume $\dim P = n$.

The name of the method comes from the following concept.

Definition 5.3 Let $V = \{v^{(1)}, \ldots, v^{(m)}\}$ be a point set in $\mathbb{R}^n$ and $\mathcal{H} = \{H_1^+, \ldots, H_k^+\}$ a set of affine half spaces in $\mathbb{R}^n$. The pair $(V, \mathcal{H})$ is called a *double description* of a polytope P if we have

$$P = \operatorname{conv} V = H_1^+ \cap \cdots \cap H_k^+.$$

Exercise 5.4 How should the term 'double description' be extended to arbitrary polyhedra?

Let $P = H_1^+ \cap \cdots \cap H_m^+$, and write $P_k := H_1^+ \cap \cdots \cap H_k^+$. Up to a projective transformation (and renumeration) we can assume that P_{n+1} is an n-simplex (see Exercise 3.57). The $n + 1$ vertices of P_{n+1} are precisely the intersections of each set of

---

Algorithm 5.2: A basic algorithm to compute the double description

---

Input: A set of affine half-spaces $\mathcal{H} = \{H_1^+, \ldots, H_m^+\}$ in $\mathbb{R}^n$, such that
$P = H_1^+ \cap \cdots \cap H_m^+$ is bounded and full-dimensional and
$P_{n+1} = H_1^+ \cap \cdots \cap H_{n+1}^+$ is an n-simplex.
Output: Point set V with conv $V = P$

1 $V_{n+1} \leftarrow$ set of vertices of P_{n+1}
2 **for** $k \leftarrow n+2, \ldots, m$ **do**
3 $\quad\lfloor$ Construct V_k with conv $V_k = P_k = P_{k-1} \cap H_k^+$ as in Lemma 5.2.

4 **return** V_m

---

n hyperplanes from $H_1, \ldots, H_{n+1}$. We can now inductively assume that we have already computed a $\mathcal{V}$-representation of $P_k = \text{conv}\{v^{(1)}, \ldots, v^{(k)}\}$. Using Lemma 5.2 we obtain Algorithm 5.2.

This basic version of the algorithm is already more efficient than the trivial method described at the beginning of this chapter. However, we can still improve it with some simple steps. Note that we have $|V_k| \leq |V_{k-1}|^2$, i.e., the number of points might be squared in each step. The improvement that we introduce below does not completely avoid this "explosion" but it does have a positive effect by avoiding redundant computations, particularly when dealing with actual applications.

The point sets V_k which are iteratively generated in Algorithm 5.2 are in general too large, since they can contain points which are not vertices. Only the vertices are necessary for a $\mathcal{V}$-representation of a polytope. A possible improvement on this method would be to set up a linear program which at each step reduces the point set V_k to the set of vertices of P_k. This technique was previously used in Exercise 4.28.

However, we would like to avoid solving additional linear programs. The above mentioned refinement relies on the observation that vertices of P_k which are not vertices of P_{k-1} are generated by intersections of edges of P_{k-1} with the new hyperplane H_k. This fact was used in the proof of Lemma 5.2. Once we know which pairs of vertices in V_{k-1} generate edges of P_{k-1}, we will only have to test those particular vertices.

For $W \subseteq V$ let

$$\mathcal{H}(W) = \left\{ H : H = \partial H^+ \text{ for an } H^+ \in \mathcal{H} \text{ and } W \subseteq H \right\}$$

be the set of supporting hyperplanes from $\mathcal{H}$ that contain all points of W. We abbreviate this as $\mathcal{H}(v, w) := \mathcal{H}(\{v, w\})$.

Lemma 5.5 *Let $(V, \mathcal{H})$ be a double description of an n-polytope $P \subseteq \mathbb{R}^n$. Given two distinct points $v, w \in V$, the set $\text{aff}\{v, w\} \cap P$ is an edge of P if and only if the affine subspace $G := \bigcap \mathcal{H}(v, w)$ is one-dimensional. In this case $\text{aff}\{v, w\} = G$ holds. Furthermore, if v and w are vertices then $\text{conv}\{v, w\} = P \cap G$.*

Proof Observe that aff$\{v, w\} \subseteq G = \bigcap \mathcal{H}(v, w)$. This is obvious for non-empty $\mathcal{H}(v, w)$. Otherwise we fix here the convention $\bigcap \emptyset = \mathbb{R}^n$.

First, let $e = \text{aff}\{v, w\} \cap P$ be an edge of P. The affine hull of each face F of P is the intersection of those hyperplanes which define the facets of P that contain F. Since $(V, \mathcal{H})$ is a double description of P, the set $\mathcal{H}$ contains all affine hyperplanes that define facets of P. In addition, every affine hyperplane that contains v and w also contains the edge e. This implies that aff$\{v, w\}$ is the intersection G of all supporting hyperplanes (from $\mathcal{H}$) that contain v and w.

For the reverse direction, let $\dim G = 1$, i.e., aff$\{v, w\} = G$. In Theorem 3.6 we showed that the faces of faces of P are faces of P themselves. This implies that every intersection of supporting hyperplanes with P defines a face of P. In particular this holds for $G \cap P$ and the assumption about the dimension implies $\dim(G \cap P) \leq 1$. Since the points v and w of G were chosen to be distinct points of P we have that $G \cap P = e$ is an edge. $\qquad\qquad\Box$

To fully realize the advantages resulting from this lemma, we have to study how to make the double description $(V, \mathcal{H})$ accessible as a data structure. We also want to extend the convex hull problem in such a way that we can handle $\mathcal{H}$-descriptions of unbounded (fully-dimensional) pointed polyhedra. Handling non-pointed polyhedra is the task of Exercise 5.13. We showed in Chapter 3 that a polyhedron is pointed if and only if it is projectively equivalent to a polytope. As usual we use homogeneous coordinates. Geometrically, the transformation to homogeneous coordinates can be interpreted as follows. Instead of working with pointed polyhedra $P \subseteq \mathbb{R}^n$, we work with the polyhedral cones which are generated by P:

$$Q = \left\{(\lambda, \lambda x) : x \in P, \ \lambda \geq 0\right\} \subseteq \mathbb{R}^{n+1}.$$

The vertices and rays of P, which we originally wanted to compute, correspond to the uniquely defined minimal generating system of Q as a positive hull: Let $V, R \subseteq \mathbb{R}^n$ be given with

$$P = \text{conv } V + \text{pos } R$$

as in Exercise 3.41. Then we have

$$Q = \text{pos}\left(\{(1, v) : v \in V\} \cup \{(0, r) : r \in R\}\right).$$

In the following let

$$W = \left\{w^{(1)}, \ldots, w^{(m)}\right\} := \left\{(1, v) : v \in V\right\} \cup \left\{(0, r) : r \in R\right\} \subseteq \mathbb{R}^{n+1}$$

be a positive generating system of the cone Q. To be able to distinguish P from its *homogenization Q*, we will refer to the elements of W as *vectors*. Through the homogenization, affine half-spaces in $\mathbb{R}^n$ become *linear half-spaces* in $\mathbb{R}^{n+1}$, i.e., affine half-spaces which contain the origin in $\mathbb{R}^{n+1}$. E.g., a simplex in $\mathbb{R}^n$ generates a *simplicial cone* in $\mathbb{R}^{n+1}$. The polytope edges, which played the key role in Lemma 5.5, correspond precisely to the two-dimensional faces of the homogenization.

The following is a useful way to represent the data: The coordinates of vectors from $W = \{w^{(1)}, \ldots, w^{(m)}\}$ are saved as columns of an $(n+1) \times m$-matrix which we will also call W. The linear half-spaces $\mathcal{H} = \{H_1^+, \ldots, H_k^+\}$ are represented by their coordinate vectors $h^{(1)}, \ldots, h^{(k)} \in (\mathbb{R}^{n+1})^*$ where we assume $H_i^+ = \{x : h^{(i)}x \geq 0\}$. By analogy to the vectors, we use $\mathcal{H}$ as the symbol for the $k \times (n+1)$-matrix consisting of the row vectors $h^{(1)}, \ldots, h^{(k)}$. We use the following homogeneous version of the incidence matrix of Section 3.6 and Exercise 3.55.

Definition 5.6 Let $(W, \mathcal{H})$ be the double description of a pointed cone $Q \subseteq \mathbb{R}^{n+1}$ with $W \in \mathbb{R}^{(n+1) \times m}$ and $\mathcal{H} \in \mathbb{R}^{k \times (n+1)}$. The matrix $I(W, \mathcal{H}) \in \{0, 1\}^{k \times m}$ with $I(W, \mathcal{H}) = (I_{ij})$ defined by

$$I_{ij} = \begin{cases} 1 & \text{if } w^{(j)} \in H_i = \partial H_i^+, \text{ i.e., } h^{(i)} w^{(j)} = 0, \\ 0 & \text{otherwise} \end{cases}$$

is called the *incidence matrix* of $(W, \mathcal{H})$.

The rows of the incidence matrix $I := I(W, \mathcal{H})$ of the cone Q can be interpreted as the characteristic functions of the set of vectors from W which lie on the corresponding hyperplane. Analogously, the columns of I correspond to sets of supporting hyperplanes which contain a fixed vector from W. In this way we can determine the set $\mathcal{H}(w^{(r)}, w^{(s)})$ from Lemma 5.5 as the intersection of two sets which are given by characteristic functions; many programming languages allow for the efficient implementation of this as a bit-wise "and". This allows us to identify the set $\mathcal{H}(w^{(r)}, w^{(s)})$ with the submatrix consisting of those rows of the matrix $\mathcal{H}$ which have a 1 in their r-th and s-th column. The dimension of the intersection of all supporting hyperplanes which contain $w^{(r)}$ and $w^{(s)}$ is therefore $n + 1$ minus the rank of the submatrix $\mathcal{H}(w^{(r)}, w^{(s)})$.

The natural formulation of the crucial Lemma 5.5 shows that it is most convenient to study the double description method in the homogeneous setting. Putting the pieces together, as shown in Algorithm 5.3, we can compute a minimal positive generating system of a polyhedral cone in $\mathbb{R}^{n+1}$ defined by linear inequalities. This is slightly more general than computing convex hulls.

We conclude this section with a detailed description of an example of the functionality of the loop in Steps 8 to 11 of Algorithm 5.3.

Example 5.7 Let $n = 3$ and

$$\mathcal{H} = \begin{pmatrix} 1 & -1 & -1 & 0 \\ 1 & -1 & 2 & 0 \\ 1 & 2 & -1 & 0 \\ 1 & 0 & 0 & 1 \\ 2 & -1 & -1 & -1 \end{pmatrix} \in \mathbb{R}^{5 \times 4}.$$

One can easily verify that the cone $Q = \{x \in \mathbb{R}^4 : \mathcal{H}x \geq 0\}$ is full-dimensional, since the ray $\mathbb{R}_{\geq 0}(1, 0, 0, 0)^T$ passes through the interior. Furthermore, we have that $Q_4 =$

Algorithm 5.3: An algorithm for the double description in homogeneous form

---

Input: Matrix $\mathcal{H} \in \mathbb{R}^{k \times (n+1)}$ with row vectors $h^{(1)}, \ldots, h^{(k)}$ such that
  $Q = \{x \in \mathbb{R}^{n+1} : \mathcal{H}x \geq 0\}$ is a full-dimensional pointed cone and
  $Q_{n+1} := \{x \in \mathbb{R}^{n+1} : h^{(1)}x \geq 0, \ldots, h^{(n+1)}x \geq 0\}$ is a simplicial cone.

Output: Set W of vectors with pos $W = Q$

1 Let $W_{n+1} \in \mathbb{R}^{(n+1) \times (n+1)}$ be a matrix whose columns positively generate Q_{n+1}.

2 **for** $i \leftarrow n+2, \ldots, k$ **do**

3   Create W_{i-1}^{+} from those columns of W_{i-1} that lie on the positive side of $h^{(i)}$ and create W_{i-1}^{-} from the columns on the negative side.

4   **if** $W_{i-1}^{-} = \emptyset$ **then**

5        $W_i \leftarrow W_{i-1}$

6   **else**

7        $X \leftarrow \emptyset$

8   **foreach** Pair (w, w') of columns of W_{i-1}^{+} and W_{i-1}^{-} **do**

9   **if** rank $\mathcal{H}_{i-1}(w, w') = n - 1$ **then**

10   Choose x as generator of the kernel of the matrix $\mathcal{H}'_{i-1}(w, w')$ that consists of the rows of $\mathcal{H}_{i-1}(w, w')$ and $h^{(i)}$.

11              $X \leftarrow X \cup \{x\}$

12   Let W_i be the matrix consisting of the columns of W_{i-1} without the columns of W_{i-1}^{-} and enhanced by the column vectors from X.

13 **return** W_k

$\{x \in \mathbb{R}^4 : h^{(1)}x \geq 0, \ldots, h^{(4)}x \geq 0\}$ is a simplicial cone whose rays correspond to the columns of the following matrix

$$W_4 = \begin{pmatrix} 1 & 1 & 1 & 0 \\ 1 & 0 & -1 & 0 \\ 0 & 1 & -1 & 0 \\ -1 & -1 & -1 & 1 \end{pmatrix} \in \mathbb{R}^{4 \times 4}.$$

Now the fifth and last row of the matrix $\mathcal{H}$ defines the subsets W_4^{+} (consisting of the first three columns of W_4) and W_4^{-} (last column of W_4). The incidence matrix of the double description is then the following:

$$I(W_4, \mathcal{H}_4) = \begin{pmatrix} 1 & 1 & 0 & 1 \\ 1 & 0 & 1 & 1 \\ 0 & 1 & 1 & 1 \\ 1 & 1 & 1 & 0 \end{pmatrix}.$$

As an example we study the pair of rays $(w^{(1)}, w^{(4)}) \in W_4^+ \times W_4^-$. By the definition of the incidence matrix, the first two vectors $h^{(1)}$, $h^{(2)}$ satisfy $h^{(j)}w^{(1)} = 0$ and $h^{(j)}w^{(4)} = 0$ (for $j = 1, 2$). This gives

$$\mathcal{H}_4(w^{(1)}, w^{(4)}) = \begin{pmatrix} 1 & -1 & -1 & 0 \\ 1 & -1 & 2 & 0 \end{pmatrix}$$

and

$$\mathcal{H}'_4(w^{(1)}, w^{(4)}) = \begin{pmatrix} 1 & -1 & -1 & 0 \\ 1 & -1 & 2 & 0 \\ 2 & -1 & -1 & -1 \end{pmatrix}.$$

The matrix $\mathcal{H}_4(w^{(1)}, w^{(4)})$ clearly has rank 2, which implies that $\mathrm{pos}\{w^{(1)}, w^{(4)}\}$ is a face of the cone Q_4 of dimension $4 - 2 = 2 = n - 1$. The vector $(1, 1, 0, 1)^T$ spans the kernel of $\mathcal{H}'_4(w^{(1)}, w^{(4)})$. Analogous computations for the pairs $(w^{(2)}, w^{(4)})$ and $(w^{(3)}, w^{(4)})$ yield two more columns. Putting this together we arrive at

$$W = W_5 = \begin{pmatrix} 1 & 1 & 1 & 1 & 1 & 1 \\ 1 & 0 & -1 & 1 & 0 & -1 \\ 0 & 1 & -1 & 0 & 1 & -1 \\ -1 & -1 & -1 & 1 & 1 & 4 \end{pmatrix}.$$

If we now *dehomogenize*, i.e., we intersect $Q = \mathrm{pos}\, W = \{(x_0, x_1, x_2, x_3)^T \in \mathbb{R}^4 : \mathcal{H}x \geq 0\}$ with the affine hyperplane in $\mathbb{R}^4$ defined by $x_0 = 1$, we obtain a simple 3-polytope with five facets which is combinatorially equivalent to a prism over a triangle. The rows of $\mathcal{H}$ and the columns of W describe homogeneous coordinates of facets and vertices of P respectively. The incidence matrix defined by the vertices and facets of P coincides with the incidence matrix of the double description $(W, \mathcal{H})$ of the cone Q:

$$I(W, \mathcal{H}) = \begin{pmatrix} 1 & 1 & 0 & 1 & 1 & 0 \\ 1 & 0 & 1 & 1 & 0 & 1 \\ 0 & 1 & 1 & 0 & 1 & 1 \\ 1 & 1 & 1 & 0 & 0 & 0 \\ 0 & 0 & 0 & 1 & 1 & 1 \end{pmatrix}.$$

5.3 Convex Hulls in the Plane

Two dimensional polytopes coincide with convex polygons. The edges of a convex polygon form the facets and the vertices can be ordered cyclically (clockwise or counter-clockwise). Let a finite set of points in the plane be given as columns of a matrix $M \in \mathbb{R}^{2 \times m}$. Then the planar convex hull problem is the computation of a list of column indices that defines such a cyclic ordering of the vertices. Depending on the context, it may be necessary to choose one of the two orientations or to fix a specific point as the starting vertex.

Algorithm 5.4: The `Divide-and-Conquer` method for computing convex hulls in the plane

---

Input: Finite point set $V = \{v^{(1)}, \ldots, v^{(m)}\} \subseteq \mathbb{R}^2$
Output: Vertices of conv V in cyclic order

1   **if** $m \leq 2$ **then**
2   **return** V
3   **else**
4   Sort V by the first coordinate.
5   Divide V in two disjoint sets L and R, where L contains the left $\lfloor m/2 \rfloor$ and R the right $\lceil m/2 \rceil$ points of V.
6   Recursively compute conv L and conv R.
7   Compute $\text{conv}(L \cup R)$ from conv L and conv R.

Note that degenerate cases of lower dimensional polytopes in $\mathbb{R}^2$ can be coded by such a list as well, which then contains only one index (if the dimension is 0), or two indices (if the dimension is 1). To keep the language simple, we shall call these degenerate polytopes *polygons*.

We now introduce an algorithm of Preparata and Hong [84], which relies on the commonly used "divide-and-conquer" principle of computer science. The basic idea is to divide the original problem into many sub-problems, solve these smaller problems recursively and combine the sub-solutions, thus forming a solution to the original problem. A classic example of this principle is the `MergeSort` sorting algorithm described in Appendix C.1.

To simplify the presentation of Preparata and Hong's algorithm we make an extra assumption. In the exercises at the end of this chapter we will see how to extend the algorithm to the general case. In contrast to the convention used elsewhere in this book, we say that a point set $V \subseteq \mathbb{R}^2$ is in *general position* if no three points are colinear and every *vertical* $[a : -1 : 0]$, for $a \in \mathbb{R}$, contains at most one point in V.

In Algorithm 5.4, the actual computational problem is of course hidden in the last step, where we have to compute the common convex hull of two polygons which are given as a cyclic list of vertices. Our assumption that no two points lie on the same vertical simplifies the situation since this implies that conv L and conv R are disjoint and that there exists a dividing vertical line. The central observation here is that in this situation those vertices of $\text{conv}(L \cup R)$ which are vertices of L (or R) are ordered successively by the cyclic ordering.

One consequence of L and R being vertically separated is that there exist four common supporting lines to L and R; see Fig. 5.1. As with smooth convex sets, we call these common supporting lines *double tangents*. Exactly two of these four double tangents define facets of the common convex hull of L and R. Since L and R are vertically separated we can talk about the *upper* and *lower* double tangent. Computing the common convex hull of L and R is therefore equivalent to computing the upper and lower double tangent of two vertically separated polygons. In addition,

Fig. 5.1 The four double tangents to two disjoint polygons

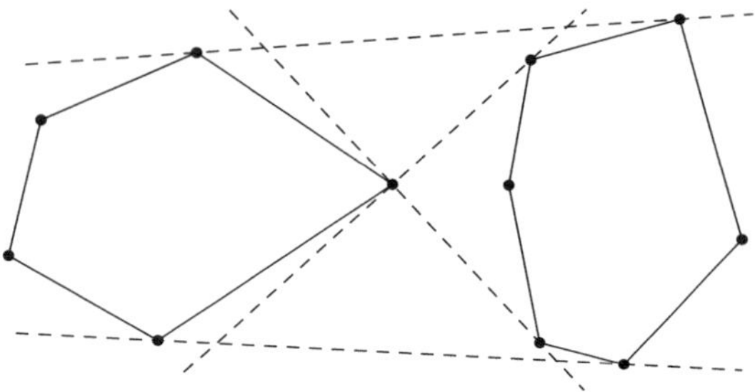

the problem of computing the upper double tangent and the problem of computing the lower double tangent are equivalent since we can obtain the upper double tangent of L and R by computing the lower double tangent of $(-R, -L)$. Now we obtain an algorithm to compute the convex hull in the plane by combining the following algorithm with the divide-and-conquer method.

It remains to be checked if the Lower-Double-Tangent algorithm is correct. This is not obvious since it has to be shown that the outer loop terminates. To do this we need a further definition and a preliminary lemma.

Each pair of vertices v and w of a polygon defines two polygonal arcs, one in which v appears before w and one where v appears after w with respect to the counter-clockwise cyclic order of the polygon's vertices. For a polygon in general position, the left-most vertex and the right-most vertex define the *upper* and the *lower half*.

Lemma 5.8 *The lower double tangent to two vertically separated polygons L and R intersects both L and R in the lower half.*

Proof The lower half comprises precisely those facets whose outer normal points down. The outer normal of a supporting line to L (or R) which points down lies in the cone of normals of the facets of the lower half. $\qquad\square$

Since the algorithm progresses cyclically in a fixed direction on both polygons, its termination is a consequence of the following statement. In some sense the interiors of L and R "block" the algorithm after finitely many steps.

Lemma 5.9 *There is no step of the algorithm where the segment $[v^{(i)}, w^{(k)}]$ could intersect the interior of L or of R.*

Proof In the beginning this condition is satisfied by construction. To show that the condition is satisfied in subsequent steps we use induction. Assume that $[v^{(i)}, w^{(k)}]$ does not intersect the interior of L and R. Using symmetry arguments we can also assume that i will be decreased in the next step. That is, we assume that $[v^{(i)}, w^{(k)}]$ is not a lower supporting line to L. Then $v^{(i-1)}$ lies below the line $\mathrm{aff}\{v^{(i)}, w^{(k)}\}$ and $[v^{(i-1)}, w^{(k)}]$ does not intersect the interior of L. $\qquad\square$

Algorithm 5.5: `Lower-Double-Tangent(L,R)`

Input: Two finite polygons $L = (v^{(0)}, \ldots, v^{(l-1)})$ and $R = (w^{(0)}, \ldots, w^{(r-1)})$, given as a list of their vertices in cyclic order counterclockwise, such that there exists a separating vertical line where L lies on the left and R on the right side

Output: Lower double tangent T

1 $v^{(i)} \leftarrow$ right-most vertex of L

2 $w^{(k)} \leftarrow$ left-most vertex of R

3 **while** $T \leftarrow \text{aff}\{v^{(i)}, w^{(k)}\}$ *is not lower double tangent* **do**

4   **while** T *is not a lower supporting line to L* **do**

5          $i \leftarrow i - 1 \bmod l$

6   **while** T *is not a lower supporting line to R* **do**

7          $k \leftarrow k + 1 \bmod r$

8 **return** T

We would like to determine the complexity of the divide-and-conquer algorithm in its worst case. This will be done in a way that is typical for algorithms of this type. When regarding the input size, we neglect the point coordinates for which, for all geometric primitives, the same unit costs occur. The complexity of the algorithm `Lower-Double-Tangent` (Algorithm 5.5) is clearly $O(l + r)$. If we denote the complexity of `Divide-and-Conquer` by $C(m)$ we have the recursion $C(2m) = 2C(m) + O(m)$. First, we assume that the number of input points $m = 2^b$ is a power of 2. Every division step will then divide the point set into two sets of exactly the same size. Then we obtain

$$
\begin{aligned}
C(m) &= C\left(2^b\right) \\
&= 2C\left(2^{b-1}\right) + O\left(2^b\right) \\
&= 2\left(C\left(2^{b-2}\right) + O\left(2^{b-1}\right)\right) + O\left(2^b\right) = 2C\left(2^{b-2}\right) + 2O\left(2^b\right) \\
&= 2C\left(2^{b-3}\right) + 3O\left(2^b\right) = \cdots = bO\left(2^b\right) = O(m \log m).
\end{aligned}
$$

If m is not a power of 2, then the smallest power of 2 that is larger than m is at most twice as large as m. The complexity analysis above changes only by a multiplicative constant which is suppressed in the O-notation. We summarize this with the following theorem.

Theorem 5.10 *The algorithm* `Divide-and-Conquer` *computes the convex hull of m points in $\mathbb{R}^2$ with complexity $O(m \log m)$.*

5.4 Inspection Using `polymake`

`polymake` offers several convex hull algorithms, some of them via interfaces to other software, others as part of the `polymake` system. The double description algorithm is the standard algorithm. Internally, `polymake` calls `cddlib` [43].

We will start with the $\mathcal{V}$-description of a polytope. In contrast to the previous chapter where we entered the coordinates manually, we now use `polymake`'s standard constructions. The function `cube` with the single argument "3" generates the standard cube $[-1, 1]^3$.

```
polytope > $C3=cube(3);
```

The following function `edge_middle` takes the cube $C3 as input, computes its edge mid-points and defines a new polytope as the convex hull of these. The task of Exercise 5.16 is to show that the edge mid-points are always the vertices of the new polytope.

```
$P=edge_middle($C3);
```

The new object $P comes with a range of properties which are already known.

```
polytope > print join " ", $P->list_properties();
VERTICES BOUNDED FEASIBLE
```

Each of them can be printed or used for further computations.

```
polytope > print $P->VERTICES;
1 0 -1 -1
1 -1 0 -1
1 1 0 -1
1 0 1 -1
1 -1 -1 0
1 1 -1 0
1 0 -1 1
1 -1 1 0
1 -1 0 1
1 1 1 0
1 1 0 1
1 0 1 1
```

```
polytope > print $P->BOUNDED, " ", $P->FEASIBLE;
1 1
```

The property VERTICES lists the vertices of the polytope in homogeneous coordinates. The boolean properties BOUNDED and FEASIBLE indicate that $P is a bounded polyhedron, i.e., a polytope, which is not empty. Performing a convex hull computation is now as easy as printing the FACETS.

```
polytope > print $P->FACETS;
1 0 0 -1
2 -1 1 -1
1 0 1 0
```

```
2  1 -1  1
1  1  0  0
2  1  1  1
2  1  1 -1
2  1 -1 -1
2 -1  1  1
1  0  0  1
2 -1 -1  1
2 -1 -1 -1
1  0 -1  0
1 -1  0  0
```

The polytope in $P is actually a *cuboctahedron* which is one of the *Archimedean solids*; see Fig. 5.2.

5.5 Exercises

Exercise 5.11 Let

$$P = \mathrm{conv}\{v^{(1)}, \ldots, v^{(m)}\} = H_1^+ \cap \cdots \cap H_l^+ \subseteq \mathbb{R}^n$$

be an n-polytope in double description with pairwise distinct half-spaces H_1^+, $\ldots, H_l^+$, and let $V_i := \{v^{(j)} \in H_i : 1 \leq j \leq m\}$ be the set of given points which lie on the hyperplane H_i. Show that H_i^+ is redundant if and only if there exists an index $k \in \{1, \ldots, l\}$ such that $V_i \subsetneq V_k$.

Exercise 5.12 Let $(V, \mathcal{H})$ be a double description of an $(n+1)$-polytope P and let $\pi : \mathbb{R}^{n+1} \to \mathbb{R}^n$ be the linear projection to the first n coordinates. Exercise 3.58 shows that the image $\pi(P)$ is also a polytope. Compute a double description of $\pi(P)$.

Throughout the double description algorithm, the step-wise intersections with hyperplanes become iterated projections to coordinate subspaces in the polar form. In its dual form, this method corresponds to *Fourier–Motzkin-Elimination*. Exercise 5.12 illustrates one elimination step.

Exercise 5.13 How can we alter Algorithm 5.2 so that it also works for non-pointed polyhedra?

Exercise 5.14 How can we alter the divide-and-conquer algorithm from Section 5.3 to compute the area of a polygon that is defined by its vertices?

Exercise 5.15 How can we alter the divide-and-conquer algorithm so that it computes the convex hull of a point set that is not in general position?

Fig. 5.2 The cuboctahedron

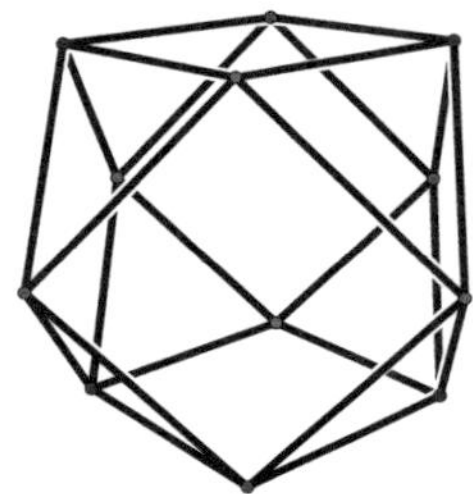

Exercise 5.16 Let P be an arbitrary polytope with vertex set $\{v^{(1)}, \ldots, v^{(m)}\} \subseteq \mathbb{R}^n$ and edge set

$$\left\{ \left[v^{(i)}, v^{(j)} \right] : (i, j) \in I \right\}$$

for an appropriate set $I \subseteq \{1, \ldots, m\} \times \{1, \ldots, m\}$. Show that the set of *edge midpoints*

$$W := \left\{ \frac{1}{2} \left(v^{(i)} + v^{(j)} \right) : (i, j) \in I \right\}$$

is the vertex set of the polytope conv W.

Figure 5.2 shows an example of the construction in Exercise 5.16 where P is the standard 3-cube.

5.6 Remarks

The double description algorithm which has briefly been introduced here is used in practical applications and is particularly useful for relatively high-dimensional non-simple polytopes. A detailed description can be found in Fukuda and Prodon [44].

The m vertices of an n-polytope defined by ℓ (facet defining) affine half-spaces can be computed in $O(\ell m n)$ time using the "reverse search" method of Avis and Fukuda [8]; reverse search works for non-simple polytopes as well, but in that setting is often inferior to the double description method; see Avis, Bremner and Seidel [7].

A further class of convex hull algorithms computes from the given point set, in addition to the facets of the convex hull, a triangulation. An example of this class is "beneath-and-beyond"; see Edelsbrunner [38, §8.4] and Joswig [67].

The divide-and-conquer principle can be extended and sometimes yields asymptotically optimal algorithms for lower dimensions. In dimension 2 and 3 one can obtain $O(m \log \ell)$-algorithms; see Clarkson and Shor [23] and Chan [19]. Chan, Snoeyink and Yap [20] describe an $O((m + \ell) \log^2 \ell)$-algorithm to compute the ℓ facets of a 4-polytope defined by m points.

The Upper-bound Theorem limits the number of facets of an n-polytope with m vertices to $\binom{m}{\lfloor n/2 \rfloor}$. When we fix the dimension n as a constant, then $\binom{m}{\lfloor n/2 \rfloor} \in$

$O(m^{\lfloor n/2 \rfloor})$ has a polynomial bound. Chazelle [21] was able to provide an algorithm which, for constant dimension, is in the worst case asymptotically optimal and has a run time of order $O(m \log m + m^{\lfloor n/2 \rfloor})$.

An interesting quality measurement for convex hull algorithms of arbitrary dimension can be obtained when we measure the run time with respect to the combination of input and output size. This is known as the *combined run-time* of a convex hull algorithm. It is unknown if there exists an algorithm that has a polynomially bounded combined run-time which computes the convex hull. Khachiyan et al. [69] recently showed that it is #P-hard (in combined run-time) to enumerate all vertices of an unbounded polyhedron which is given by inequalities. But this result does not imply that it is #P-hard (in combined run-time) to enumerate all vertices and additionally all rays. Therefore the complexity of enumerating all vertices of a polytope is still unknown.

Chapter 6
Voronoi Diagrams

Let S be a finite point set in $\mathbb{R}^n$. Since S is compact, for every point $x \in \mathbb{R}^n$ there exists a closest point in S (which is not necessarily unique) with respect to the Euclidean norm $\| \cdot \|$. The set of all points in $\mathbb{R}^n$ that have a fixed point $s \in S$ as their nearest "neighbor" is a polyhedron. This mapping induces a decomposition of $\mathbb{R}^n$ into polyhedral "regions", the Voronoi diagram of S. Numerous applications of computational geometry begin with the computation of a Voronoi diagram.

We will first study the geometry of single Voronoi regions. To be able to discuss the arrangement of all Voronoi regions, we will introduce the general concept of a polyhedral complex. The main result of this chapter is the relationship between Voronoi diagrams and the convex hull problem from the previous chapter. We conclude the chapter by discussing an algorithm for the computation of Voronoi diagrams in the plane and its application to the post-office problem from the introduction.

6.1 Voronoi Regions

In this chapter, $S \subseteq \mathbb{R}^n$ always denotes a finite point set in $\mathbb{R}^n$ and $\| \cdot \|$ is the Euclidean norm. The Euclidean distance between two points $x, y \in \mathbb{R}^n$ is denoted by

$$\operatorname{dist}(x, y) := \|x - y\| = \sqrt{\langle x - y, x - y \rangle}.$$

For each point $s \in S$ we define the *Voronoi region*

$$\mathrm{VR}_S(s) := \left\{ x \in \mathbb{R}^n : \operatorname{dist}(x, s) \leq \operatorname{dist}(x, q) \text{ for all } q \in S \right\}$$

as the set of points in $\mathbb{R}^n$ for which s is the nearest point from S. In this case, s is called a *nearest neighbor* (with respect to S).

Example 6.1 We study the case where $S = \{s, t\} \subseteq \mathbb{R}^n$ consists of exactly two distinct points. The set

$$h(s, t) := \left\{ x \in \mathbb{R}^n : \operatorname{dist}(x, s) = \operatorname{dist}(x, t) \right\} = \mathrm{VR}_{\{s,t\}}(s) \cap \mathrm{VR}_{\{s,t\}}(t)$$

M. Joswig, T. Theobald, *Polyhedral and Algebraic Methods in Computational Geometry*, 81
Universitext, DOI 10.1007/978-1-4471-4817-3_6,
© Springer-Verlag London 2013

consisting of those points which have both s and t as a nearest neighbor is an affine hyperplane: We have

$$\langle x - s, x - s \rangle - \langle x - t, x - t \rangle = \sum_{i=1}^{n}(x_i - s_i)^2 - \sum_{i=1}^{n}(x_i - t_i)^2$$

$$= \sum_{i=1}^{n} 2(t_i - s_i)x_i + \sum_{i=1}^{n}(s_i^2 - t_i^2),$$

which implies that x is contained in $h(s, t)$ if and only if

$$\left(\sum_{i=1}^{n}(s_i^2 - t_i^2), 2(t_1 - s_1), \ldots, 2(t_n - s_n)\right)(1, x_1, \ldots, x_n)^T = 0. \qquad (6.1)$$

In other words, the set $h(s, t) = \mathrm{VR}_{\{s,t\}}(s) \cap \mathrm{VR}_{\{s,t\}}(t)$ is precisely the affine hyperplane in $\mathbb{R}^n$ which has the homogeneous coordinates

$$\left[\sum_{i=1}^{n}(s_i^2 - t_i^2) : 2(t_1 - s_1) : \cdots : 2(t_n - s_n)\right]. \qquad (6.2)$$

The Voronoi regions of s and t are the affine half-spaces which are defined by this hyperplane. We always define the orientation of $h(s, t)$ as in (6.2). Thus, we have $\mathrm{VR}_{\{s,t\}}(s) = h(s, t)^-$ and $\mathrm{VR}_{\{s,t\}}(t) = h(s, t)^+$. The vectors $s - t$ and $t - s$ are normal to the hyperplane $h(s, t)$ which (weakly) separates the two Voronoi regions.

The above observations about Voronoi regions of a two-element point set lead to the following statement.

Proposition 6.2 *Let $S \subseteq \mathbb{R}^n$ be finite. For $s \in S$ we have*

$$\mathrm{VR}_S(s) = \bigcap_{t \in S \setminus \{s\}} \mathrm{VR}_{\{s,t\}}(s) = \bigcap_{t \in S \setminus \{s\}} h(s, t)^-.$$

In particular, each Voronoi region is a (not necessarily bounded) polyhedron with at most $|S| - 1$ facets.

Exercise 6.3 Give conditions which imply that all Voronoi regions are pointed polyhedra.

Exercise 6.4 Show that a point $s \in S$ lies on the boundary of the convex hull $\mathrm{conv}\, S$ if and only if its Voronoi region $\mathrm{VR}_S(s)$ is unbounded.

6.2 Polyhedral Complexes

We know from the previous section that the Voronoi regions of a finite point set in $\mathbb{R}^n$ are polyhedra. By construction, it is clear that these polyhedra cover the whole space $\mathbb{R}^n$. However, this alone does not reveal all of the important structural properties of Voronoi regions.

Definition 6.5 A *polyhedral complex* $\mathcal{C}$ is a finite set of polyhedra in $\mathbb{R}^n$ which satisfies the following conditions.

(a) $\emptyset \in \mathcal{C}$;
(b) If $P \in \mathcal{C}$, then all faces of P are also contained in $\mathcal{C}$;
(c) The intersection $P \cap Q$ of two polyhedra $P, Q \in \mathcal{C}$ is a (possibly empty) face of P and of Q.

The third condition is sometimes called the *intersection condition*. The elements of $\mathcal{C}$ are called *faces* and the *dimension* of $\mathcal{C}$ is the highest dimension of a face of $\mathcal{C}$. A polyhedral complex whose faces are polytopes is called a *polytopal complex*. A *simplicial complex* is a polytopal complex whose faces are simplices.

For a polyhedral complex $\mathcal{C}$ in $\mathbb{R}^n$ let

$$|\mathcal{C}| := \bigcup_{F \in \mathcal{C}} F \subseteq \mathbb{R}^n$$

be the *set covered by* $\mathcal{C}$. A *polyhedral* (respectively *polytopal* or *simplicial*) *decomposition* of a set $M \subseteq \mathbb{R}^n$ is a polyhedral (respectively polytopal or simplicial) complex $\mathcal{C}$ such that $|\mathcal{C}| = M$. A simplicial decomposition is also called a *triangulation*.

Example 6.6 Let $P \subseteq \mathbb{R}^n$ be an n-polyhedron. Then the face lattice $\mathcal{F}(P)$ is an n-dimensional polyhedral complex. The set of all proper faces defines an $(n-1)$-dimensional polyhedral complex that covers the boundary ∂P. This second complex is called the *boundary complex* of P.

The faces of a polyhedral complex $\mathcal{C}$ are partially ordered by inclusion; this is the *face poset* of $\mathcal{C}$. This notion agrees with the face lattice of a polytope if we view that polytope as a trivial polytopal complex as in the previous example.

Let $\mathcal{V}(S)$ be the set of all Voronoi regions of a finite set $S \subseteq \mathbb{R}^n$.

Theorem 6.7 *The set $\mathcal{V}(S)$ satisfies the intersection condition.*

Proof Let $s, t \in S$ be two distinct points. We can assume that the intersection

$$F := \mathrm{VR}_S(s) \cap \mathrm{VR}_S(t)$$

is non-empty. Proposition 6.2 states that $\mathrm{VR}_S(s) \subseteq h(s,t)^-$ and that $\mathrm{VR}_S(t) \subseteq h(s,t)^+$. This implies that $F \subseteq h(s,t)^- \cap h(s,t)^+ = h(s,t)$. Since we assumed

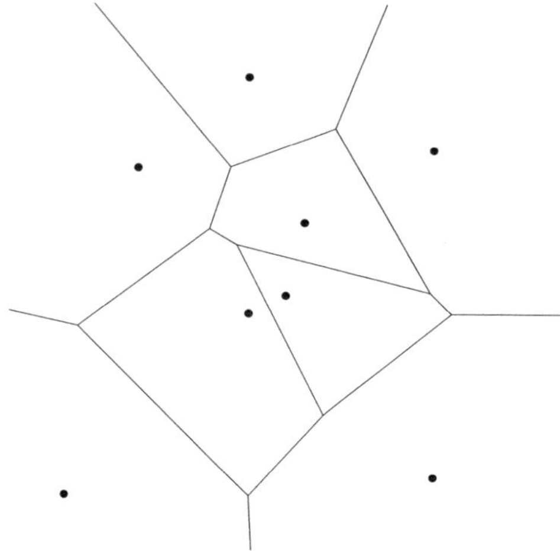

Fig. 6.1 The Voronoi diagram of a point set in the plane

$F \neq \emptyset$, we know that $h(s, t)$ is a supporting hyperplane of $\mathrm{VR}_S(s)$ and also of $\mathrm{VR}_S(t)$. Thus, $F = \mathrm{VR}_S(s) \cap \mathrm{VR}_S(t) = \mathrm{VR}_S(s) \cap \mathrm{VR}_S(t) \cap h(s, t)$ is a non-empty face of both Voronoi regions. $\square$

Every non-empty finite set C of polyhedra in $\mathbb{R}^n$ that satisfies the intersection condition *generates* a polyhedral complex

$$[C] := \{F : F \text{ is the face of a polyhedron in } C\}.$$

The previous theorem motivates the following definition.

Definition 6.8 The polyhedral complex

$$\mathrm{VD}(S) := \big[\{\mathrm{VR}_S(s) : s \in S\}\big]$$

is called the *Voronoi diagram* of a finite set $S \subseteq \mathbb{R}^n$.

The faces of a Voronoi diagram are called *Voronoi cells*. The Voronoi regions are the maximal Voronoi cells (with respect to inclusion or dimension). Figure 6.1 depicts an example of a Voronoi diagram of a point set in the plane.

Remark 6.9 The definition of f-vectors can be extended to arbitrary polyhedral complexes.

6.3 Voronoi Diagrams and Convex Hulls

As we will see in the following chapters, Voronoi diagrams play a key role in several applications. Many interesting algorithms, e.g., the curve reconstruction algorithm

`NN-Crust` from Chapter 11 below, have the computation of a Voronoi diagram as their very first step. This motivates the questions of how a Voronoi diagram should be computed and what a suitable data structure would be for Voronoi diagrams.

A first observation is that convex hull algorithms are useful for the computation of Voronoi diagrams: Every region is given as a polyhedron in the $\mathcal{H}$-description. For m given points in $\mathbb{R}^n$ we obtain, by computing m dual convex hulls in $\mathbb{R}^n$, a $\mathcal{V}$-description of all Voronoi regions. Regardless of the efficiency of this method, the main disadvantage of it is that it does not directly provide a description of the relative position of the different Voronoi regions to one another. The main result of this chapter is the statement that a Voronoi diagram in $\mathbb{R}^n$ is a projection of an unbounded polyhedron in $\mathbb{R}^{n+1}$. Specifically, this reduces the construction of a Voronoi diagram to a single convex hull problem in $\mathbb{R}^{n+1}$.

To clarify the notation, we will embed $\mathbb{R}^n$ in $\mathbb{R}^{n+1}$ by adding the coordinate x_{n+1}. In particular, we will sometimes denote a point in $\mathbb{R}^{n+1}$ by (x, x_{n+1}) for $x \in \mathbb{R}^n$ and $x_{n+1} \in \mathbb{R}$.

Let

$$U := \left\{ x \in \mathbb{R}^{n+1} : x_{n+1} = x_1^2 + x_2^2 + \cdots + x_n^2 \right\} \tag{6.3}$$

be the *standard paraboloid* in $\mathbb{R}^{n+1}$. For a point $p \in \mathbb{R}^n$ let $T(p)$ denote the tangent hyperplane to the paraboloid U at $p_U := (p, \|p\|^2)$.

Lemma 6.10 *For every point $p \in \mathbb{R}^n$ we have*

$$T(p) = \left[-\|p\|^2 : 2p_1 : \cdots : 2p_n : -1 \right].$$

Proof We know from calculus that the tangent hyperplane to the graph of a differentiable function $u : \mathbb{R}^n \to \mathbb{R}$ at a point $(p, u(p))$ can be described by the linear equation

$$x_{n+1} = u(p) + \langle u'(p), x - p \rangle$$

(see, e.g., [73]). In our case, we have $u(p) = p_1^2 + p_2^2 + \cdots + p_n^2 = \|p\|^2$, and thus the gradient satisfies $u'(p) = (2p_1, \ldots, 2p_n) = 2p$. Substituting yields

$$x_{n+1} = \|p\|^2 + \langle 2p, x - p \rangle = -\|p\|^2 + 2\langle p, x \rangle,$$

and thus we obtain the desired representation of the tangent hyperplane in homogeneous coordinates. $\square$

In the following, we imagine that the x_{n+1}-direction of the coordinate system points vertically upwards.

Lemma 6.11 *Let $p, x \in \mathbb{R}^n$ and $x_U = (x, \|x\|^2)$ be the point lying above x on U. Then x_U lies above $T(p)$, i.e., in the affine half-space $T(p)^+$ with respect to the homogeneous coordinates from Lemma 6.10. The vertical distance from x_U to $T(p)$ is $\|x - p\|^2$.*

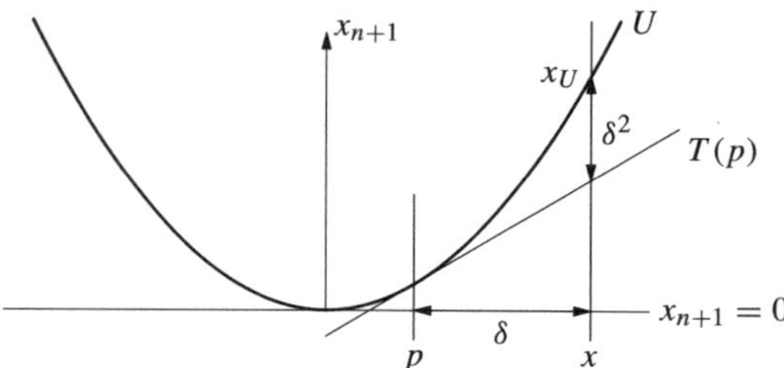

Fig. 6.2 The distance computation for $n = 1$. Due to the rotation invariance of U, the 2-dimensional figure suggests the proper intuition for higher dimensions. Here, we have $\delta = \|x - p\|$

Proof The x_{n+1}-coordinate of x_U is $\sum_{i=1}^{n} x_i^2$, and by Lemma 6.10 the x_{n+1}-coordinate of the point on the hyperplane $T(p)$ above x is

$$2p_1 x_1 + \cdots + 2p_n x_n - p_1^2 - \cdots - p_n^2.$$

The distance from x_U to $T(p)$ is $(x_1 - p_1)^2 + \cdots + (x_n - p_n)^2 = \|x - p\|^2$. Figure 6.2 illustrates this computation. $\qquad\square$

Let S be an m-element subset of $\mathbb{R}^n$. For a point $s \in S$ we have that $T(s)^+$ is the affine half-space above the tangent hyperplane at U.

Due to the monotonicity of the function $\delta \mapsto \delta^2$ on the positive half-line, we can interpret Proposition 6.2 using Lemma 6.11 in the following way: A point $x \in \mathbb{R}^n$ is contained in the Voronoi region $\mathrm{VR}_S(s)$ if and only if for all $T(t)$, where $t \in S$, the hyperplane $T(s)$ is the one that has the smallest vertical distance from the point x_U. This implies the following statement; see Fig. 6.3.

Theorem 6.12 *The Voronoi diagram of S is the orthogonal projection of the boundary complex of the polyhedron $\mathcal{P}(S) := \bigcap_{s \in S} T(s)^+$ to the hyperplane $x_{n+1} = 0$.*

Corollary 6.13 *The total number of cells of a Voronoi diagram of an m-element point set in $\mathbb{R}^n$ is of order $O(m^{\lceil n/2 \rceil})$.*

Proof The total number of cells of a Voronoi diagram can be bounded by the maximal number of faces of an $\mathcal{H}$-polyhedron with m facets in $\mathbb{R}^{n+1}$. The dual version of the asymptotic Upper-bound Theorem, Theorem 3.46, therefore implies that the total number of faces is of order $O(m^{\lceil n/2 \rceil})$, since $\lfloor (n+1)/2 \rfloor = \lceil n/2 \rceil$. $\qquad\square$

Theorem 6.12 specifically states that the space is partitioned by the relative interior of the cells of $\mathrm{VD}(S)$. For an arbitrary point $x \in \mathbb{R}^n$ let

$$\mathbb{B}_S(x) := \left\{ y \in \mathbb{R}^n : \mathrm{dist}(x, y) < \mathrm{dist}(x, s) \text{ for all } s \in S \right\} \qquad (6.4)$$

be the largest open ball with center x which does not contain a point of S. Furthermore, let

$$S(x) := \partial \mathbb{B}_S(x) \cap S.$$

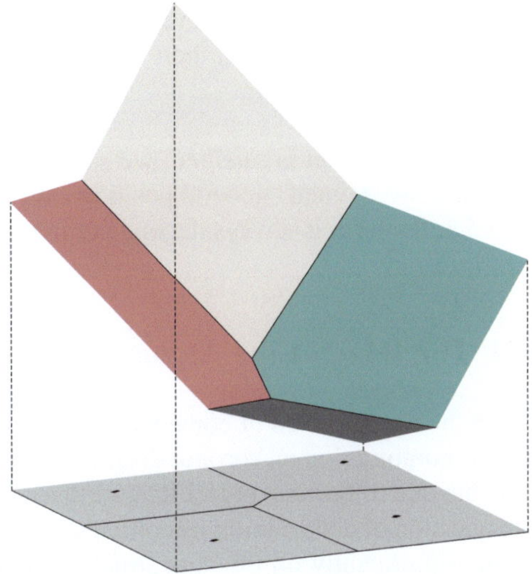

Fig. 6.3 A Voronoi diagram obtained by an orthogonal projection

Theorem 6.14 *The uniquely determined relatively open cell of* VD(S) *that contains a given point* $x \in \mathbb{R}^n$ *has dimension* $n - \dim \operatorname{aff} S(x)$.

Proof The point x is contained in a relatively open k-cell C of VD(S) if and only if there exists a series of facets $F_1, \ldots, F_{n-k+1}$ of the polyhedron $\bigcap_{s \in S} T(s)^+$ for which:

$$G_1 \supsetneq G_2 \supsetneq \cdots \supsetneq G_{n-k+1} =: G, \tag{6.5}$$

where $G_i := F_1 \cap \cdots \cap F_i$ and C is the orthogonal projection of G to $\mathbb{R}^n$; see Exercise 3.59. The decreasing chain condition in (6.5) is satisfied for the facets $F_1, \ldots, F_{n-k+1}$ if and only if $G = F_1 \cap \cdots \cap F_{n-k+1}$ is non-empty and the normals of the facets are linearly independent.

By Lemma 6.10 we have that $(2s_1, \ldots, 2s_n, -1)^T$ is a normal vector to the facet $T(s)$ for $s \in S$. Therefore, the normal vectors to the facets corresponding to a subset $S' \subseteq S$ are linearly independent if and only if the points of S' are affinely independent. Altogether, this proves the statement. $\qquad\square$

In the next section we will focus on the planar case $n = 2$. Therefore, we are interested in the following special cases of Theorem 6.14.

Corollary 6.15 *Let* $S \subseteq \mathbb{R}^2$ *be finite.*

(a) *A point* $x \in \mathbb{R}^2$ *is a vertex of the Voronoi diagram* VD(S) *if and only if* $S(x)$ *contains at least three points.*
(b) *A point* $x \in \mathbb{R}^2$ *lies in the relative interior of an edge of* VD(S) *if and only if* $S(x)$ *consists of exactly two points.*

For a vertex x of the Voronoi diagram VD(S) we call the ball $\mathbb{B}_S(x)$ from (6.4) the *Voronoi disk* around x. The boundary $\partial \mathbb{B}_S(x)$ is called the *Voronoi circle*.

Exercise 6.16 Show that if every $(n + 2)$-element subset of $S \subseteq \mathbb{R}^n$ does not lie on a common $(n - 1)$-sphere, then the lifted polyhedron is simple and therefore every Voronoi region is simple.

If this condition is satisfied, we say that the points in S are in *general position*. Note that we defined "general position" slightly differently in Chapter 3 and in Section 5.3; the term is always dependent on the context.

6.4 The Beach Line Algorithm

As in the computation of convex hulls in Section 5.3, there exist special algorithms for the computation of Voronoi diagrams in the planar case. We introduce here an algorithm due to Fortune [42]. First, we discuss the geometric idea, and then approach the question of determining its complexity. In this particular case, the complexity depends significantly on the data structures employed. With respect to this property, this algorithm is an exception within this text.

Fortune's beach line algorithm is a so-called *sweep line method*. The idea is to construct the Voronoi diagram of a finite point set $S \subseteq \mathbb{R}^2$ step-by-step. Here, we can imagine the vertical axis as a time-scale that is traversed from top to bottom. In this interpretation, at a certain time τ only a part of the Voronoi diagram has been revealed by the algorithm. For a point s from the input set S we then have that s is known at time τ if $s_2 \geq \tau$. The horizontal line $H_\tau = [-\tau : 0 : 1]$ is the sweep line for time τ and the affine half-space $[-\tau : 0 : 1]^+$ contains the previously detected points from S. The next natural question is which part of the Voronoi diagram is actually known at time τ.

The set of points in $\mathbb{R}^2$ that have the same distance from a point p and a (non-incident) line G is a parabola, which we denote here by $\mathrm{Par}(p, G)$ (see Exercise 6.18 below). For every point $s \in S$ with $s_2 > \tau$ which is known at time τ, all points which are closer to s than to any possible unknown point of S lie above the parabola $\mathrm{Par}(s, H_t)$. The term "above" makes sense here since the symmetry axis of $\mathrm{Par}(s, H_\tau)$ is parallel to the vertical axis. The time τ is called *generic* if $H_\tau \cap S = \emptyset$. If we denote the points on or above the parabola by $\mathrm{Par}(s, H_\tau)^+$, then, according to our notation for affine half-spaces, we get the following lemma.

Lemma 6.17 *The part of the Voronoi diagram which is known at time τ is contained in the set*

$$\bigcup_{s \in S} \mathrm{Par}(s, H_\tau)^+$$

for each generic time $\tau \in \mathbb{R}$.

If τ is generic, the set $\bigcup_{s \in S} \mathrm{Par}(s, H_\tau)^+$ is homeomorphic to an affine half-space. Its boundary B_τ is a union of parabolic arcs that resembles the appearance of waves approaching a beach; see Fig. 6.4. This is the reason why the boundary curve is called the "beach line", and this term gives the algorithm its name. Note that each vertical line intersects the *beach line* B_τ in exactly one point; this property is inherited from the individual parabolas.

Exercise 6.18 Determine a parametrization of the parabola $\mathrm{Par}(s, H_\tau)$ for a given s and $\tau \in \mathbb{R}$. That is, search for $a, b, c \in \mathbb{R}$ such that

$$\mathrm{Par}(s, H_\tau) = \left\{ \begin{pmatrix} x \\ ax^2 + bx + c \end{pmatrix} : x \in \mathbb{R} \right\},$$

subject to the condition that $s_2 > \tau$.

A point $s \in S$ with the property that $\mathrm{Par}(s, H_\tau)$ is part of the beach line is said to be *active* at time τ.

Now we briefly discuss what happens at a non-generic time τ. For sufficiently small $\epsilon > 0$ we have that $\tau - \epsilon$ is a generic time. The smaller ϵ is, the steeper the parabola $\mathrm{Par}(s, H_{\tau-\epsilon})$ will be. This is rigorously formulated in the following exercise.

Exercise 6.19 Let $s = (s_1, s_2)^T \in S$ be a point with $\tau = s_2$. Show that

$$\lim_{\epsilon \to 0^+} \mathrm{Par}(s, H_{\tau-\epsilon}) = \left\{ \begin{pmatrix} s_1 \\ \sigma \end{pmatrix} \in \mathbb{R}^2 : \sigma \geq s_2 \right\}.$$

Here, we mean convergence with respect to the Hausdorff metric. How is it possible to use this to define the beach line for non-generic times? [*Hint*: Look at Snapshot 2 in Fig. 6.4.]

Lemma 6.20 *If τ is generic, then each parabolic arc in $\mathrm{Par}(s, H_\tau) \cap B_\tau$, for $s \in S$, is contained in the corresponding Voronoi region $\mathrm{VR}_S(s)$.*

The set $\mathrm{Par}(s, H_\tau) \cap B_\tau$ may consist of several parabolic arcs, e.g., snapshot 2 in Fig. 6.4. Here the parabolic arc for b is divided as soon as the point d becomes known, i.e., at time d_2.

Proof For $x \in \mathrm{Par}(s, H_\tau) \cap B_\tau$ let $\delta := \mathrm{dist}(x, s) = \mathrm{dist}(x, H_\tau)$ and assume $x \notin \mathrm{VR}_S(s)$. By Corollary 6.15 the open disk B around x with radius δ contains a point $r \in S$. Since $B \subseteq H_\tau^+$, we have that r is known at time τ. But x is above the parabola $\mathrm{Par}(r, H_\tau)$, which contradicts x being contained in the beach line B_τ. $\quad\square$

The next question is to determine how the beach line changes as the time τ changes (in the direction of smaller values). Here, of course, the relevant times are those when a certain point $s = (s_1, s_2)^T \in S$ is first detected; see Snapshot 2 in Fig. 6.4. This time s_2 will be called a *point event*. It is a consequence of Lemma 6.20, and of the convexity of the Voronoi regions, that new parabolic arcs can only arise at point events; the beach line cannot be pierced from behind by a parabola. For a generic τ we have that, by construction, the beach line has only finitely many points where it is not differentiable, since it is the union of finitely many parabolic arcs; these points are called *breakpoints*.

Lemma 6.21 *If τ is generic then every breakpoint of B_τ lies on an edge of the Voronoi diagram.*

Proof Let x be a breakpoint of the beach line B_τ at time τ. Then there exist two active points $r, s \in S$ with $x \in \mathrm{Par}(r, H_\tau) \cap \mathrm{Par}(s, H_\tau)$ and the statement follows from Lemma 6.20. $\square$

We assume that the vertical line $[-s_1 : 1 : 0]$ through s intersects the beach line B_{s_2} at a point x which is contained in a unique parabolic arc $\mathrm{Par}(r, H_{s_2})$; here $r \in S$ is an active point. By construction we have that $x \in \mathrm{VR}_S(r) \cap \mathrm{VR}_S(s)$ and $\mathrm{VR}_S(r) \cap \mathrm{VR}_S(s)$ is an edge of the Voronoi diagram; this edge is detected (partly) for the first time at time s_2. For a sufficiently small $\epsilon > 0$, a part of the parabola $\mathrm{Par}(s, H_{s_2-\epsilon})$ lies on the beach line, say with the breakpoints x and y. Then, the segment $[x, y]$ is the intersection of the Voronoi edge $\mathrm{VR}_S(r) \cap \mathrm{VR}_S(s)$ and the set above the beach line. Thus, new edges are discovered at point events.

By Corollary 6.15, every vertex v of $\mathrm{VR}(S)$ lies on a circle through at least three points of S. The point in time at which a circle through at least three points from S is detected is called a *circle event*. In other words, we have a circle event at time τ if the sweep line H_τ is the lower tangent to a circle through at least three points of S. By Corollary 6.15 only those circle events create vertices whose circular disks have no points of S in their interior.

Now we can examine how a parabolic arc γ vanishes from the beach line. Let γ' and γ'' be, respectively, the left and the right neighbor of γ in the beach line. Let $s, s', s'' \in S$ be the points corresponding to these three parabolic arcs. We assume now that the parabolic arc γ vanishes at time τ. At the slightly later generic point in time $\tau - \epsilon$, γ' and γ'' are neighbors in the beach line. Hence, by Lemma 6.21, we know that the Voronoi regions $\mathrm{VR}_S(s')$ and $\mathrm{VR}_S(s'')$ are neighbors in $\mathrm{VD}(S)$. At time τ, γ contracts to a point v. By construction, we have that $\delta := \mathrm{dist}(v, s) = \mathrm{dist}(v, s') = \mathrm{dist}(v, s'')$ and that v is a Voronoi vertex. Also, the distance between v and the sweep line is δ at time τ. This means that τ is a circle event for the triple of points (s, s', s''). This is illustrated in Snapshots 4 and 7 in Fig. 6.4.

Exercise 6.22 Show that there are at most $2|S| - 2$ breakpoints in the beach line B_τ for a generic time τ.

Data Structures The way in which geometric data is stored is crucial for the run-time analysis of the beach line algorithm. Here we only outline the most important ideas and refer the reader to the original work of Fortune [42], and to the books [31] and [71], for more details of the implementation.

First, we have to decide in which way we want to store the output, i.e., the Voronoi diagram of a point set in the plane. One special feature of the planar case is

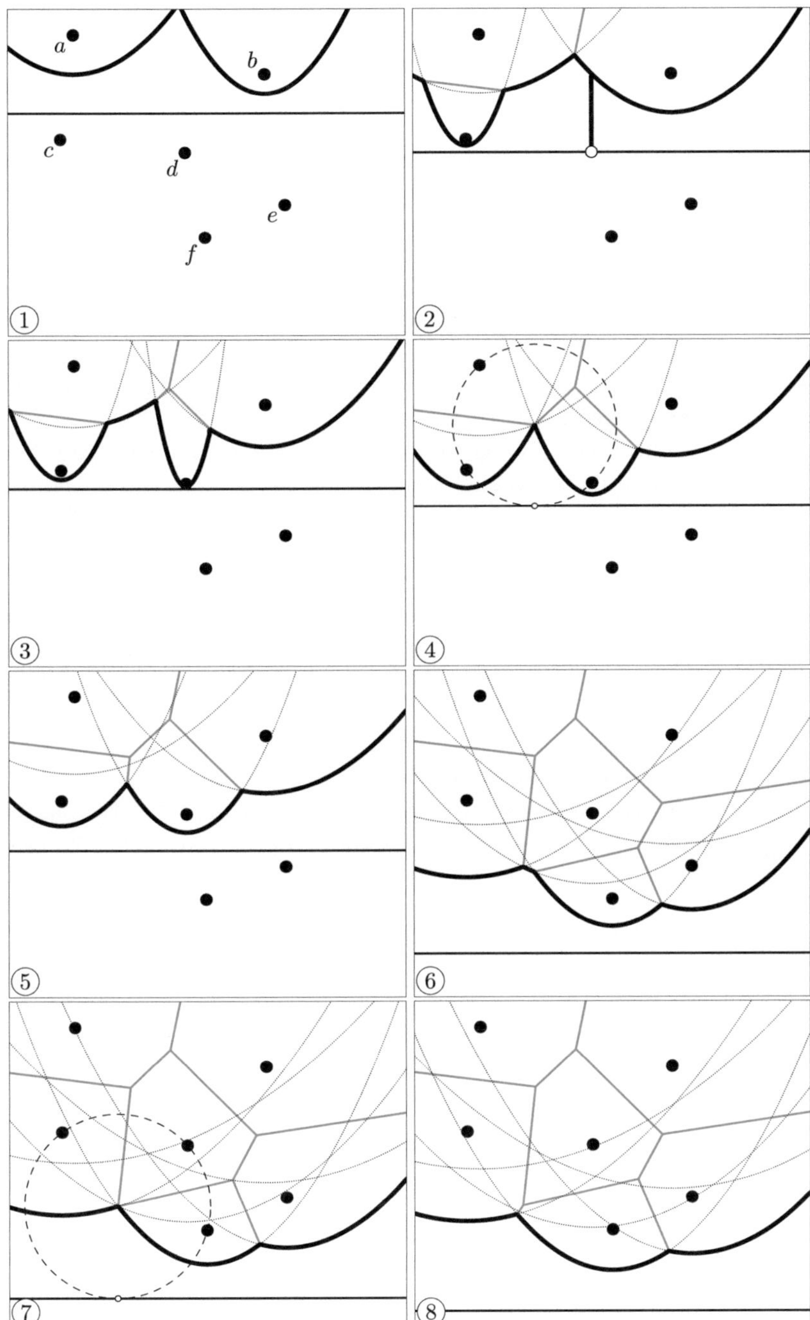

Fig. 6.4 Eight snapshots of the beach line algorithm

that we can restrict ourselves to the Voronoi edges. Every Voronoi region is a (not necessarily bounded) polygon whose edges can be cyclically ordered. Every edge is contained in exactly two regions. Therefore, if we store each edge twice with its vertices and orientation, then the regions are implicitly given by the sequence of their edges. Thus, each oriented edge stands for an incident pair of a Voronoi edge and a Voronoi region, and the Voronoi vertices are implicitly given as the endpoints of the edges.

Depending on the specific construction of the data structure, it can be problematic that some Voronoi regions are unbounded, and thus the cyclic sequence of edges does not form a complete circle. However, this can be easily addressed by using the ideal points of lines on which the unbounded edges lie as artificial Voronoi vertices. These ideal points can then be connected by artificial Voronoi edges on the ideal line so that every Voronoi region can be represented as a closed circle of (original or artificial) Voronoi edges.

In practical applications, it is common to use points on a sufficiently large bounding box, rather than artificial Voronoi vertices on the ideal line. This bounding box should be large enough to contain all points of S and all vertices of VD(S).

The data structure itself is then a *doubly linked list* of oriented edges, which are also called *half-edges*, such that each edge is stored with its two endpoints and with a reference to the next half-edge in the cyclic order. Furthermore, we store a reference to the parallel half-edge, i.e., the same edge with the opposite orientation. This data structure is also known as the *half-edge data structure*. We refer to [27, §10.2] for the implementation of doubly linked lists.

Note also that the half-edge data structure is useful for storing arbitrary planar graphs and arbitrary cell decompositions of oriented surfaces.

Before we study the beach line algorithm in detail, we have to determine a suitable way in which to code the beach line itself. Here, it is not necessary to trace the exact trajectory of each parabolic arc. We need only store the combinatorial information, i.e., the number of parabolic arcs in the beach line, the points of S to which they correspond, and the order in which they occur.

Example 6.23 The beach line from Fig. 6.5 can be coded, for example, by the ordered sequence of points $(s^{(1)}, s^{(2)}, s^{(3)}, s^{(4)}, s^{(5)})$ and the breakpoints correspond to neighboring pairs of points.

Some points can occur multiple times. For example, we see that the beach line in Snapshot 2 of Fig. 6.4, which appears shortly after a point event, can be written as (a, c, a, b, d, b).

However, coding the beach line as an ordered list is not beneficial for the run-time complexity. It is better to use a *binary search tree*. The leaves of this search tree contain points from S that each correspond to one parabolic arc on the beach line. An interior vertex stands for a breakpoint (r, s) if r is the biggest leaf in the left subtree, and s the smallest in the right subtree; see Fig. 6.5. In particular, we have that here, in contrast to the list description, the breakpoints are explicitly represented.

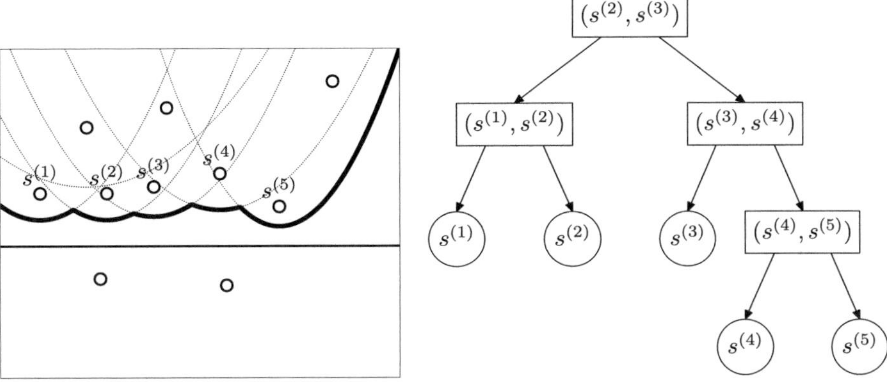

Fig. 6.5 A beach line consisting of five parabolic arcs and a representation as a search tree

For further details on the implementation of the search tree representation of the beach line, we refer to the book [31]. General binary search trees are described in [27, §12].

The search tree structure of the beach line is not sufficient to guarantee a good run-time of the algorithm. We also need the height of the search tree to be of size $O(\log m)$, where $m = |S|$, at every step of the algorithm. A search tree with this property is said to be *balanced*. Note that the coding length of the beach line, i.e., the number of parabolic arcs and breakpoints, is linear in m; see Exercise 6.22. Hence, it is possible to add or delete parabolic arcs in $O(\log m)$ time.

Various concepts are associated with the balance of search trees, for example, the so-called "red black trees" [27, §13].

Lastly, we need to establish a data structure for the point and circle events. An important aspect here is the (time-wise) order, which would suggest a list, or a search tree as a suitable representation. However, here it is crucial to immediately see the next event at every step, without having to perform a search. Therefore, a search tree is not suitable. It is also important to be able to quickly add new events at the right position in the sorted order. Thus, a list is also not suitable. The solution is a *heap*, which allows us to immediately see the next event (i.e., in constant time), and to delete this event after processing it in logarithmic time. Furthermore, we need to be able to guarantee that arbitrary new events can be added in $O(\log m)$ time. An example of a suitable data structure is a *binomial heap* [27, §19].

6.4.1 The Algorithm

Using the data structures described above, we can now detail the actual algorithm. Let $\mathcal{B}$ be a balanced search tree that represents the beach line. The queue Q stores unprocessed events, which are listed in order of their appearance. Every event in the queue Q is represented by point coordinates. The sweep line is only implicitly represented by the next event at a given time.

Algorithm 6.1: The beach line algorithm

Input: Finite point set $S \subseteq \mathbb{R}^2$
Output: VD(S) in half-edge model

```
1  B ← ∅
2  Initialize Q with all point events from S.
3  while Q ≠ ∅ do
4  │    e ← next event in Q; remove e from Q
5  │    if e point event for s ∈ S then
6  │    │    Handle-Point-Event(s, Q, B)
7  │    else
8  │    │    Handle-Circle-Event(e, Q, B)
```

The order defined by the queue, and thus the heap structure, corresponds precisely to the ordering of the events by their y-coordinate. Since the sweep line is moving from top to bottom, points with a large y-coordinate represent early events. Here, a point event corresponding to $s \in S$ is coded by the point s itself. A circle event is represented by the lowest point of the circle; when the sweep line reaches the lowest point of the circle, the whole circular disk is visible.

For a correct implementation it is crucial that the events in Q are not stored in an isolated way. It is necessary to be able to distinguish between point and circle events. Moreover, it is also useful that a point that represents a circle event also refers to the points of S that define the circle. There are a few additional references of this kind between the data structures B and Q, but we restrict ourselves to the presentation of the crucial ideas. We mainly ignore the processing of the actual Voronoi diagram in the half-edge model in our pseudo-code. This has the consequence that Algorithm 6.1 lists VD(S) as output, but we never state a return value in the code.

For our analysis, we first assume that the points in S are in general position, i.e., at most three on one circle at a time. The case where this condition is not satisfied is discussed at the end of this section.

Before we discuss the two subroutines to process point and circle events on p. 95, we will estimate the complexity of the steps of the main program. Initializing the heap Q has time complexity $O(m \log m)$ (this can be reduced to $O(m)$ when a suitable implementation is used), since there are exactly m point events. Estimating the number of possible circle events is more difficult, since there may be circle events that do not lead to Voronoi vertices. An analysis of Steps 10 to 12 of the subroutine `Handle-Point-Event` shows that every Voronoi edge can trigger at most two (potential) circle events. Therefore, by Corollary 6.13, there are at most $O(m)$ events in total; Steps 3 to 8 in Algorithm 6.1 are hence performed at most $O(m)$ times. If Q is realized as a binomial heap, it takes $O(\log m)$ time to delete an event from Q. We will show below that each point and each circle event only requires logarithmic time. This implies that the total time complexity of the beach line algorithm is $O(m \log m)$.

Every parabolic arc γ is implicitly coded in the search tree $\mathcal{B}$ as a triple $[(r,s), s, (s,t)]$, where $r, s, t \in S$ are as in Fig. 6.5. The pairs of points (r,s) and (s,t) represent the breakpoints that bound the parabolic arc. In particular, we have that the parabolic arc on the left side of γ corresponds to r and the one on the right corresponds to t.

When checking the correctness of this subroutine, note that each circle event is matched to the lowest point of the corresponding Voronoi circle. Therefore, the point events that correspond to points on a Voronoi circle, i.e., that trigger a circle event, are always correctly processed at a time prior to the circle event. This is also true for the special case where the third point of S on a Voronoi circle is simultaneously the lowest point, i.e., when a circle event and the corresponding point event occur at the same time. In this case, the following occurs: the parabolic arc corresponding to the lowest point is generated and immediately afterwards deleted by the simultaneously occurring circle event. In particular, Algorithm 6.1 always begins with at least three point events before the first circle event can occur.

Simultaneously occurring point events can be processed in arbitrary order. The same is true for simultaneously occurring circle events, since we assumed the points to be in general position. Thus, two circle events may occur at the same time, but at different places. Simultaneous point and circle events that are unrelated do not pose a problem. The only critical case, i.e., when a circle event is triggered by a simultaneous point event, was discussed above.

Step 3 in `Handle-Circle-Event` can be seen as the reverse of Step 8 in `Handle-Point-Event`. There, only those parabolic arcs are deleted which were previously generated by a point event. Step 11 of `Handle-Point-Event` can also trigger redundant circle events, but these are detected and deleted in Step 4 of `Handle-Circle-Event`.

---

1 **Procedure:** `Handle-Point-Event`$(s, Q, \mathcal{B})$

2 **if** $\mathcal{B} = \emptyset$ **then**

3   Add s to $\mathcal{B}$.

4 **else**

5   Let $\gamma = ((p,q), q, (q,r))$ be the parabolic arc in $\mathcal{B}$ above s.

6   **if** γ *refers to a circle event in* Q **then**

7   delete this event

8   Replace γ in $\mathcal{B}$ by the three parabolic arcs
$$\big[(p,q), q, (q,s)\big], \ \big[(q,s), s, (s,q)\big], \ \big[(s,q), q, (q,r)\big].$$

9   Generate a pair of new half-edges for the Voronoi edge $\mathrm{VR}(q) \cap \mathrm{VR}(s)$.

10   Compute the intersection point $v = (v_1, v_2)^T$ of the Voronoi edge corresponding to the parabolic arc $\gamma := ((q,s), s, (s,q))$ and the Voronoi edge corresponding to the parabolic arc on the left.

11   Add $(v_1, v_2 - \mathrm{dist}(v, s))$ as a potential circle event e to Q.

12   The parabolic arc γ contains a reference to e and vice-versa.

13   Proceed analogously to Steps 10 to 12 with the parabolic arc on the right side of γ.

1 **Procedure:** `Handle-Circle-Event`$(e, Q, \mathcal{B})$

2 Let γ be a parabolic arc that vanishes at the circle event e.
3 Remove γ from $\mathcal{B}$ and update the neighboring inner vertices.
4 Remove all circle events from Q which are referred to by γ or by one of its
   two neighbors.
5 Generate the center z of the circle corresponding to e as a new Voronoi vertex.
6 Generate a pair of new half-edges for the new breakpoint that emerges due to
   the removal of γ.
7 Store z as an endpoint of the two involved edges.
8 Link the edges to one another with respect to the half-edge model.

Example 6.24 We want to show how the point event illustrated in Snapshot 2 in
Fig. 6.4 affects the event queue Q. Before the point event corresponding to the
point d is processed, the queue contains three point events and one circle event:

$$Q = \big(d, (a, b, c), e, f\big).$$

The point event d triggers two new circle events. After this, at the generic time
$\tau = d_2 - \epsilon$, we have:

$$Q = \big((a, b, d), (a, c, d), e, f\big).$$

Later, the two circle events (a, b, d) and (a, c, d) will generate Voronoi vertices.
The circle event (a, b, c) vanishes at time d_2 (`Handle-Point-Event`, Step 7),
since we then know that d is contained in the circumcircle of a, b and c.

It remains to be discussed what occurs when the points in S are not in general
position. It is perhaps surprising that our algorithm works here with only a few modifications. Actually, we have that the beach line algorithm produces a valid Voronoi
diagram that may contain some edges of length 0. It is simple to detect and delete
these edges in linear time after the algorithm has terminated.

6.5 Determining the Nearest Neighbor

We now discuss the problem of finding the nearest neighbor, or the nearest post
office respectively, which we mentioned in the introduction. Given a finite point set
$S \subseteq \mathbb{R}^2$ and a point $p \in \mathbb{R}^2$, we want to determine the point $s \in S$ which minimizes
$\text{dist}(p, s)$. This problem has, of course, a very simple solution, i.e., we can compare
each distance from p to every point of S. If S consists of m points, this method
needs $O(m)$ steps.

But, when the configuration of the point set S is always the same and only the
point p changes with each call, a different approach may be better. If we expect

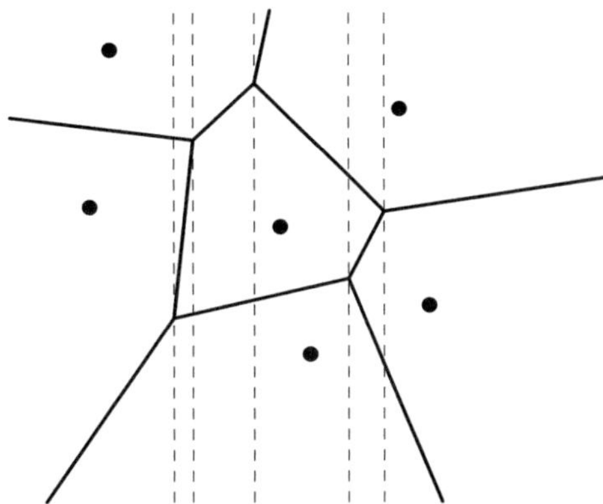

Fig. 6.6 Vertical layers in the Voronoi diagram for answering the nearest neighbor problem

many calls, it pays off to invest more time in the beginning to be able to process each later call more quickly. In the following, let m be the cardinality of S.

Our goal is to describe a data structure that enables the answer of each call in logarithmic time. To do this, we compute the Voronoi diagram of S using Fortune's beach line algorithm in $O(m \log m)$ steps.

Then, we draw a vertical line through each Voronoi vertex as depicted in Fig. 6.6. These additional lines divide the Voronoi diagram into triangles and trapezoids, and into unbounded polyhedra in the outer regions. These vertical layers are ordered from left to right. If these are stored in a balanced search tree, we can detect the layer of each point $p \in \mathbb{R}^2$ via its first coordinate p_1 in $O(\log m)$ time.

By construction we can guarantee that no vertical layer contains a vertex in its interior, so that all Voronoi edges are vertically ordered within each layer. If we also store the edges in each layer in a balanced search tree, we can detect the pair of edges that lies directly above and below p in $O(\log m)$ steps using the second coordinate p_2.

Theorem 6.25 *For an m-element point set $S \subseteq \mathbb{R}^2$ it is possible to generate a data structure in $O(m^2 \log m)$ time such that the solution to the nearest neighbor problem in S can be found in $O(\log m)$ time.*

Proof It is possible to compute the Voronoi diagram $VD(S)$ in $O(m \log m)$ time. Since there exist linearly many Voronoi vertices, there exist linearly many vertical layers. In each layer there are at most linearly many edges. In total, we have to initialize $O(m)$ balanced search trees each with $O(m)$ vertices. $\qquad\square$

6.6 Exercises

Exercise 6.26 Let S be the vertex set of the n-dimensional cross-polytope. Determine the f-vector of the Voronoi diagram $VD(S)$.

Exercise 6.27 Let $e^{(1)}, \ldots, e^{(n)}$ denote the standard basis vectors of $\mathbb{R}^n$. The vertices of the standard cube $[0, 1]^n$ are precisely the sums of pairwise distinct standard basis vectors. Show that the $n!$ simplices

$$\Delta(\sigma) := \mathrm{conv}\left\{0, e^{(\sigma(1))}, e^{(\sigma(1))} + e^{(\sigma(2))}, \ldots, e^{(\sigma(1))} + e^{(\sigma(2))} + \cdots + e^{(\sigma(n))}\right\}$$

generate a triangulation of $[0, 1]^n$, where σ runs through all elements of the symmetric group $\mathrm{Sym}\{1, \ldots, n\}$. Show that every simplex $\Delta(\sigma)$ has the same volume (i.e., $1/n!$).

Exercise 6.28 Let $m \in \mathbb{N}$ be arbitrary. Describe an m-element point set in $\mathbb{R}^2$ (in general position) for which the beach line algorithm first treats all point events and then all circle events.

6.7 Remarks

Voronoi diagrams have appeared independently over the last few centuries in different scientific disciplines. Their methodical usage in mathematics can be traced back to Dirichlet (1850) and Voronoi (1908), who used the diagrams to study quadratic forms. The presentation of a Voronoi diagram can be found as early as in Descartes' (1644) work on visualizing the mass distribution in our solar system.

Detailed discussions of this topic can be found in the books of Edelsbrunner [38], Boissonat and Yvinec [15] and de Berg et al. [31].

`polymake` computations with Voronoi diagrams will be explained in Section 7.6 below. CGAL offers a variety of methods to compute Voronoi diagrams and their generalizations, including the beach line algorithm.

Chapter 7
Delone Triangulations

We have already illustrated the utility of Voronoi diagrams with the application in Section 6.5. In fact, the neighborhood relations of points to each other which are expressed in Voronoi diagrams are used in their dual form in many other applications. This leads to the concept of Delone subdivisions (of the convex hull) of a point set. We shall discuss an application of this in Chapter 11.

As part of our study of Delone triangulations, we will explore the relation of convex hull algorithms to triangulation methods and to the computation of volumes.

7.1 Duality of Voronoi Diagrams

Let $S \subseteq \mathbb{R}^n$ be finite such that S affinely spans the space $\mathbb{R}^n$. By Theorem 6.12 we know that a Voronoi diagram $\mathrm{VD}(S)$ is generated by the vertical projection of the polyhedron $\mathcal{P}(S) = \bigcap_{s \in S} T(s)^+ \subseteq \mathbb{R}^{n+1}$ to the first n coordinates. Here, $T(s)$ denotes the tangent hyperplane of the standard paraboloid U at the point $s_U := (s, \|s\|^2)^T$, and $T(s)^+$ denotes the upper half-space. By Theorem 6.14, $\mathrm{aff}\, S = \mathbb{R}^n$ implies that $\mathcal{P}(S)$ has a vertex, i.e., it is pointed. Therefore, by Theorem 3.36, $\mathcal{P}(S)$ is projectively equivalent to a polytope. In the following we describe how to construct a polytope which is projectively equivalent to $\mathcal{P}(S)$.

To do this, we examine the projective transformation π of $\mathbb{P}^{n+1}_{\mathbb{R}}$ defined by the $(n+2) \times (n+2)$-matrix

$$
\begin{pmatrix}
1 & 0 & \cdots & \cdots & 0 & 1 \\
0 & 2 & 0 & \cdots & 0 & 0 \\
\vdots & 0 & \ddots & \ddots & \vdots & \vdots \\
\vdots & \vdots & \ddots & \ddots & 0 & \vdots \\
0 & 0 & \cdots & 0 & 2 & 0 \\
-1 & 0 & \cdots & \cdots & 0 & 1
\end{pmatrix}.
$$

As we have previously done, we regard $\mathbb{R}^{n+1}$ as a subset of $\mathbb{P}^{n+1}_{\mathbb{R}}$ via the embedding ι introduced in Section 2.1.

M. Joswig, T. Theobald, *Polyhedral and Algebraic Methods in Computational Geometry*, Universitext, DOI 10.1007/978-1-4471-4817-3_7,
© Springer-Verlag London 2013

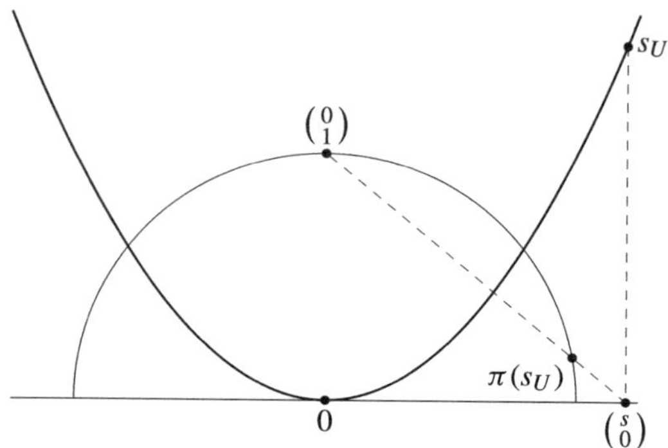

Fig. 7.1 An illustration of the standard parabola and of the map π inducing the stereographic projection

Lemma 7.1 *The projective transformation π maps the standard paraboloid $U \subseteq \mathbb{R}^{n+1}$ to the unit sphere $\mathbb{S}^n \subseteq \mathbb{R}^{n+1}$. The only point on $\mathbb{S}^n$ which is not contained in the image of U under π is the north pole $(1 : 0 : \cdots : 0 : 1)^T$. The tangential hyperplane $[1 : 0 : \cdots : 0 : 1]$ at the north pole is the image of the ideal hyperplane under π.*

Proof For a point $s \in \mathbb{R}^n$ we have

$$\pi\left(1 : s_1 : \cdots : s_n : \|s\|\right)^T = \left(1 + \|s\|^2 : 2s_1 : \cdots : 2s_n : \|s\|^2 - 1\right)^T,$$

and also $1 + \|s\|^2 > 0$. The square of the norm of the (affine) image point is

$$\left\|\left(1 + \|s\|^2 : 2s_1 : \cdots : 2s_n : \|s\|^2 - 1\right)^T\right\|^2$$

$$= \frac{4s_1^2 + \cdots + 4s_n^2 + (\|s\|^2 - 1)^2}{(1 + \|s\|^2)^2}$$

$$= 1.$$

This implies that $\pi(s)$ lies on the unit sphere.

Since π induces a stereographic projection from $\mathbb{R}^n$ to $\mathbb{S}^n \setminus \{(1 : 0 : \cdots : 0 : 1)^T\}$, we can show that $(1 : 0 : \cdots : 0 : 1)^T$ is the only point on $\mathbb{S}^n$ that is not contained in the image of π. To do this, it suffices to study the case $n = 1$. The affine point

$$\pi(s_U) = \left(\frac{2s}{1 + s^2}, \frac{s^2 - 1}{1 + s^2}\right)^T$$

is the intersection point of the unit circle and the connecting line of $(s, 0)^T$ with the north pole $(0, 1)^T$; see Fig. 7.1.

The last statement, i.e., that the ideal hyperplane $[1 : 0 : \cdots : 0]$ is mapped to the tangential hyperplane at the north pole, can be proved with a simple calculation. $\square$

Exercise 7.2 Show that the closure of the image $\overline{\pi(\mathcal{P}(S))}$ is a polytope. [*Hint*: Use Lemma 6.11 to compute a ball that contains $\pi(\mathcal{P}(S))$.]

In the following we will denote the polytope $\overline{\pi(\mathcal{P}(S))}$ by P_S. By construction, $P_S \subseteq \mathbb{R}^{n+1}$ is full-dimensional and has the origin in its interior. Its polar polytope $Q_S := P_S^\circ$ is also full-dimensional and has the origin in its interior. Since π is differentiable, Lemma 7.1 implies that all facets of P_S are tangent to $\mathbb{S}^n$. This is true for the images of the facets of $\mathcal{P}(S)$ under π, as well as for the image of the ideal hyperplane $[1 : 0 : \cdots : 0]$. This leads to the following $\mathcal{V}$-representation of Q_S:

$$Q_S = \operatorname{conv}\left(\left\{\left(1 + \|s\|^2 : 2s_1 : \cdots : 2s_n : \|s\|^2 - 1\right)^T : s \in S\right\}\right.$$

$$\left. \cup \left\{(1 : 0 : \cdots : 0 : 1)^T\right\}\right). \tag{7.1}$$

Furthermore, the points in (7.1) are the vertices of Q_S. If we apply the map π^{-1} to Q_S we obtain, since $\pi^{-1}((1 : 0 : \cdots : 0 : 1)^T) = (0 : \cdots : 0 : 1)^T$, an unbounded polyhedron

$$R_S = \operatorname{conv}\{s_U : s \in S\} + \operatorname{pos}\left\{(0, \ldots, 0, 1)^T\right\} \subseteq \mathbb{R}^{n+1}.$$

Definition 7.3 The *Delone polytope* of S,

$$\mathcal{P}^*(S) := \operatorname{conv}\{s_U : s \in S\},$$

is the convex hull of the points of S lifted to the standard paraboloid.

By construction we have that $\mathcal{P}^*(S)$ is the convex hull of the vertices of the unbounded polyhedron R_S. In Section 5.3 we defined "upper" and "lower" halves of convex polygons. We generalize this here for arbitrary polytopes.

Definition 7.4 Let h be an outer normal vector of a facet F of an $(n + 1)$-polyhedron $P \subseteq \mathbb{R}^{n+1}$. With respect to the last coordinate direction, we call F an

$$\left.\begin{array}{l} upper \\ vertical \\ lower \end{array}\right\} \text{ facet of } P \text{ if the scalar product } \langle h, e^{(n+1)}\rangle \text{ is } \left\{\begin{array}{l} > 0, \\ = 0, \\ < 0. \end{array}\right.$$

Definition 7.5 A *polytopal subdivision* of a finite point set $S \subseteq \mathbb{R}^n$ is a polytopal subdivision of the convex hull $\operatorname{conv} S$ whose vertex set consists of the points of S.

Theorem 7.6 *Let $P \subseteq \mathbb{R}^{n+1}$ be a polytope with vertex set V and let*

$$S = \left\{(v_1, \ldots, v_n)^T : v \in V\right\} \subseteq \mathbb{R}^n$$

be the projection of V to the first n coordinates. Then the lower facets of P induce a polytopal subdivision that covers the set $\operatorname{conv} S$. Furthermore, the image of every face F which is contained in a lower facet is affinely isomorphic to F. The same holds for the upper facets of P.

Proof Since the lower (and upper) facets lie in the boundary complex of P (see Example 6.6), the intersection condition is automatically satisfied. It remains to show that the projections of the lower facets of P cover $Q := \operatorname{conv} S$.

Let h be the outer normal vector of a lower facet F of P. Without loss of generality, let $\langle h, F \rangle = 0$, i.e., $\operatorname{aff} F = \operatorname{lin} F$ is a linear hyperplane. We choose a basis $(v^{(1)}, \dots, v^{(n)})$ of $\operatorname{lin} F$. Since h is perpendicular to $\operatorname{lin}\{v^{(1)}, \dots, v^{(n)}\}$, we know that $(v^{(1)}, \dots, v^{(n)}, h)$ is a basis of $\mathbb{R}^{n+1}$. Additionally, since $\langle h, e^{(n+1)} \rangle \neq 0$, the vectors $v^{(1)} - v^{(1)}_{n+1} e^{(n+1)}, \dots, v^{(n)} - v^{(n)}_{n+1} e^{(n+1)}, -e^{(n+1)}$ also form a basis. Therefore, the orthogonal projection of F is linearly (or in the general case, affinely) isomorphic to F. The same argument works for upper facets.

Since Q is a polytope, it remains to show that each vertex v of Q lies on the orthogonal projection of a lower and an upper facet. The preimage of v under the orthogonal projection is either a vertex v' or a vertical edge of P. We begin by examining the first case. Since v' is "visible" in the projection, there exists a vector h in the normal cone of v' such that $\langle h, e^{(n+1)} \rangle = 0$. And since v' is the unique preimage of v, we know that h is contained in the relative interior of the normal cone of v'. Thus, there exist vectors h_+, h_- in the normal cone of v' such that $\langle h_+, e^{(n+1)} \rangle > 0$ and $\langle h_-, e^{(n+1)} \rangle < 0$. Since $\langle h_+, e^{(n+1)} \rangle > 0$, there exists at least one upper facet that contains v'. Furthermore, $\langle h_-, e^{(n+1)} \rangle < 0$ implies that there is at least one lower facet that contains v'.

We still need to address the case where the preimage of v is a vertical edge $[v', w']$ of P. Assume, without loss of generality, that v' lies above w'. Then there exists a vector h_+ in the normal cone of v' such that $\langle h_+, e^{(n+1)} \rangle > 0$, and there exists a vector h_- in the normal cone of w' such that $\langle h_-, e^{(n+1)} \rangle < 0$. Thus, v' is contained in at least one upper facet and w' is contained in at least one lower facet of P. $\qquad\square$

The proof also shows that each polytope in $\mathbb{R}^{n+1}$ has at least one lower *and* at least one upper facet. This is not necessarily true for unbounded polyhedra.

7.2 The Delone Subdivision

We now examine the lower facets of the Delone polytope

$$\mathcal{P}^*(S) := \operatorname{conv}\{s_U : s \in S\}$$

of the finite point set S.

Theorem 7.7 *The lower facets of $\mathcal{P}^*(S)$ induce, by vertical projection, a polytopal subdivision $\mathrm{DS}(S)$ of S whose face poset is anti-isomorphic to the face poset of the Voronoi diagram $\mathrm{VD}(S)$.*

Proof The vertex set of the polytope $\mathcal{P}^*(S)$ is the set $\{s_U : s \in S\}$. Also, all facets of the polyhedron $\mathcal{P}(S)$ are lower facets. Together with Theorem 7.6, this implies that $\mathrm{DS}(S)$ is a polytopal subdivision of S.

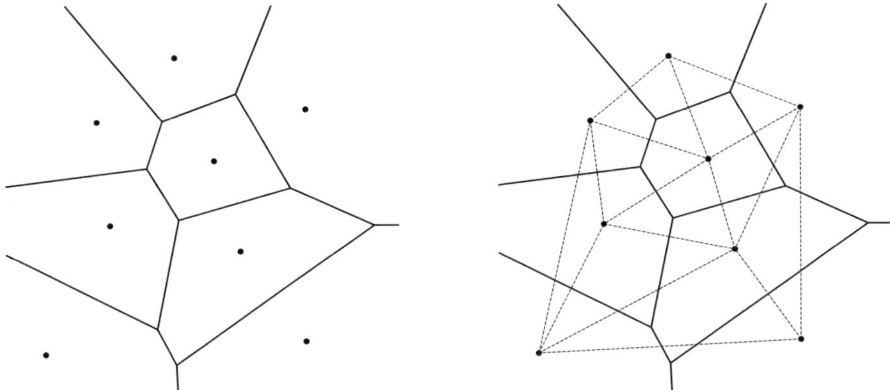

Fig. 7.2 A Voronoi Diagram and the corresponding Delone subdivision

Each Voronoi cell $F \in \mathrm{VD}(S)$ can be expressed as the intersection of Voronoi regions. This means that there exists a set of points $F(S) \subseteq S$ such that $F = \bigcap_{s \in F(S)} \mathrm{VR}_S(s)$. The map

$$\kappa : \mathrm{VD}(S) \to \mathrm{DS}(S), \quad F = \bigcap_{s \in F(S)} \mathrm{VR}_S(s) \mapsto \mathrm{conv}\, F(S) \qquad (7.2)$$

is bijective, and thus by Theorem 3.32 reverses the inclusion relation between the faces. Together, this shows that κ defines an anti-isomorphism of the face poset of $\mathrm{VD}(S)$ onto the face poset of $\mathrm{DS}(S)$. $\qquad \square$

Definition 7.8 The polytopal subdivision $\mathrm{DS}(S)$ of the set S in Theorem 7.7 is called the *Delone subdivision* of S.

In particular, Theorem 7.7 states: For $s, s' \in S$ the segment $[s, s']$ is an edge of the Delone subdivision $\mathrm{DS}(S)$ if and only if the Voronoi regions $\mathrm{VR}_S(s)$ and $\mathrm{VR}_S(s')$ have a common facet.

As in Chapter 6, we say that the points of S are in *general position* if no $(n+2)$-element subset of S lies on a common sphere.

Corollary 7.9 *If the points of S are in general position, then* $\mathrm{DS}(S)$ *is a triangulation.*

Proof The statement follows from Exercise 6.16 and Corollary 3.33. $\qquad \square$

The points in Fig. 7.2 are in general position and their Delone subdivision is a triangulation. The following definition makes use of the notion of refinement: we say that a polytopal subdivision $\mathcal{S}_1$ of S *refines* a polytopal subdivision $\mathcal{S}_2$ of S if every polytope of $\mathcal{S}_1$ is contained in some polytope of $\mathcal{S}_2$.

Definition 7.10 A *Delone triangulation* of S is a triangulation of S that refines the Delone subdivision.

If S is in general position, then $DS(S)$ is the unique Delone triangulation of S. We now discuss an important property of the Delone subdivision that results from its duality to the Voronoi diagram. As before, let $S \subseteq \mathbb{R}^n$ be finite.

Theorem 7.11 *Let $T \subseteq S$ be an arbitrary subset. The polytope $\operatorname{conv} T$ is a face of the Delone subdivision $DS(S)$ if and only if there exists an open n-dimensional ball B such that $B \cap S = \emptyset$ and $\partial B \cap S = T$.*

Proof First, let $F := \operatorname{conv} T$ be a k-face of $DS(S)$. By Theorem 7.7, F is dual to an $(n - k)$-face F^* of the Voronoi diagram $VD(S)$. Let x be a point in the relative interior of F^*. By Theorem 6.14 the largest open ball $\mathbb{B}_S(x)$ around x that does not contain a point from S satisfies the condition $\partial \mathbb{B}_S(x) \cap S = T$.

Now let B be an open ball such that $B \cap S = \emptyset$ and $\partial B \cap S = T$. The center of B lies in the intersection of the Voronoi regions that correspond to the points in T. Again, Theorems 6.14 and 7.7 imply that $\operatorname{conv} T$ is a face of $DS(S)$. $\qquad\square$

Exercise 7.12 Prove that the lower facets of $\mathcal{P}^*(S)$ are precisely the bounded faces of R_S. [*Hint*: The task of Exercise 6.4 was to show that a point $s \in S$ lies on the boundary of the convex hull $\operatorname{conv} S$ if and only if its Voronoi region $VR_S(s)$ is unbounded.]

Exercise 7.13 Let κ be the bijection from $VD(S)$ to $DS(S)$ defined in (7.2). Show that every face $F \in VD(S)$ is orthogonal to its image $\kappa(F) \in DS(S)$.

7.3 Computation of Volumes

We have already seen the versatility of convex hull algorithms when we applied them to Voronoi diagrams (and via duality to Delone subdivisions). To give the reader an idea of how central convex hull methods are to linear geometry, we will take a brief detour to discuss the computation of volumes.

Corollary 7.9 stated that the Delone subdivision of a point set S in general position is a triangulation. In this case, we can sum the volumes of the maximal simplices in $DS(S)$ to compute the volume of the convex hull $\operatorname{conv} S$; see Algorithm 7.1.

To complete the description of this method we review the computation of the volume of a simplex. Let $s^{(1)}, \ldots, s^{(n+1)} \in \mathbb{R}^n$ be points in general position, i.e., $\Delta := \operatorname{conv}\{s^{(1)}, \ldots, s^{(n+1)}\}$ is a simplex. Then,

$$\operatorname{vol} \Delta = \frac{1}{n!} \cdot \det \begin{pmatrix} 1 & 1 & \cdots & 1 \\ s_1^{(1)} & s_1^{(2)} & \cdots & s_1^{(n+1)} \\ \vdots & \vdots & \ddots & \vdots \\ s_n^{(1)} & s_n^{(2)} & \cdots & s_n^{(n+1)} \end{pmatrix} \qquad (7.3)$$

Algorithm 7.1: The volume of the convex hull of points in general position

Input: $S \subseteq \mathbb{R}^n$ finite, in general position, aff $S = \mathbb{R}^n$
Output: volume of conv S
1 compute the Delone triangulation $\mathcal{D} = \mathrm{DS}(S)$
2 $v \leftarrow 0$
3 **for** Δ maximal face in $\mathcal{D}$ **do**
4    $\quad \lfloor \quad v \leftarrow v + \mathrm{vol}\, \Delta$
5 **return** v

---

is the volume of Δ. There is a beautiful geometric proof for this. From linear algebra we know that the determinant in (7.3) (without the factor $1/n!$) is the volume of the parallelepiped spanned by the vectors $s^{(1)}, \ldots, s^{(n+1)} \in \mathbb{R}^n$. Every parallelepiped can be transformed into a cuboid via a shear mapping. Shear mappings are affine transformations which preserve volume. Hence, the above statement about the volume of Δ follows from Exercise 6.27, where we studied the triangulations of the standard cube $[0, 1]^n$. Alternatively, we can compute the volume of the simplex inductively with a calculation.

In general, of course, we cannot assume that the point set S is in general position. This is where the following exercise comes in.

Exercise 7.14 Show that each polytope P admits a triangulation whose vertices are precisely the vertices of P. [*Hint*: Use Corollary 7.9. If the vertices of P are not in general position employ the perturbation procedure from Lemma 3.48.]

Whether or not S is in general position, replacing $\mathcal{D}$ in Algorithm 7.1 by any triangulation of conv S gives an algorithm for volume computation. If S is not in general position, for instance, the triangulation obtained from Exercise 7.14 can be used.

Note that this method of computing the volume via Delone triangulations is of purely theoretical relevance. In the remarks at the end of this chapter we refer to approaches which are more relevant to practical applications.

Remark 7.15 In some practical applications it is necessary to compute the volume of non-convex geometric objects. Using the inclusion-exclusion formula, see Gallier [45, §4.4], one can generalize (exact or approximative) methods for computing the volume of convex polytopes to arbitrary finite unions of polytopes.

7.4 Optimality of Delone Triangulations

It is known that Delone triangulations (especially in the plane) satisfy several optimal properties within the set of all triangulations of a given set of points. For example, we have that in $\mathbb{R}^2$ the minimal angle appearing in the triangles is maximized (as will be shown in Corollary 7.28). In higher dimensions the situation is more complicated. We will show that the maximal radius of the circumsphere is minimized.

Let $\mathcal{T}$ be an arbitrary triangulation of a given finite point set $S \subseteq \mathbb{R}^n$ such that $\dim \mathrm{aff}\, S = n$. For every point $x \in \mathrm{conv}\, S$ there exists a (not necessarily unique) n-simplex $\Delta \in \mathcal{T}$ that contains x. Let

$$\mathbb{S}(c, \rho) := \left\{ y \in \mathbb{R}^n : \|y - c\| = \rho \right\}$$

be the unique sphere with center c and radius ρ which contains the vertices of Δ. We call $\mathbb{S}(c, \rho)$ the *sphere spanned by* Δ. We define the number $\psi_{\mathcal{T}}(x, \Delta)$ as

$$\psi_{\mathcal{T}}(x, \Delta) := \rho^2 - \|x - c\|^2.$$

Clearly $\psi_{\mathcal{T}}$ can only be non-negative. Furthermore, $\psi_{\mathcal{T}}(x, \Delta) = 0$ if and only if x lies on the sphere $\mathbb{S}(c, \rho)$, i.e., x is a vertex of $\mathcal{T}$. For a Delone triangulation the value of the function does not depend on the simplex Δ. The proof is left to the reader in the following exercise.

Exercise 7.16 Let $\mathcal{D}$ be a Delone triangulation of S. Show that for any two simplices Δ, Δ' in $\mathcal{D}$ that contain x we have

$$\psi_{\mathcal{D}}(x, \Delta) = \psi_{\mathcal{D}}(x, \Delta').$$

Therefore, we can unambiguously write $\psi_{\mathcal{D}}(x)$ instead of $\psi_{\mathcal{D}}(x, \Delta)$ for a Delone triangulation $\mathcal{D}$.

Before we study the map ψ for various triangulations of S, we need a general statement about the intersection of the standard paraboloid U from (6.3) with affine hyperplanes.

Proposition 7.17 *Let* $p \in \mathbb{R}^{n+1}$ *with* $p_{n+1} < \sum_{i=1}^n p_i^2$. *Then the intersection of the standard paraboloid U with the affine hyperplane*

$$H = \left\{ x \in \mathbb{R}^{n+1} : x_{n+1} = 2 \sum_{i=1}^n p_i x_i - p_{n+1} \right\} \tag{7.4}$$

is mapped by the vertical projection to the sphere

$$\left\{ x \in \mathbb{R}^n : \sum_{i=1}^n (x_i - p_i)^2 = \sum_{i=1}^n p_i^2 - p_{n+1} \right\} \subseteq \mathbb{R}^n. \tag{7.5}$$

Conversely, the map $x \mapsto x_U = (x, \|x\|^2)$ lifts every sphere in $\mathbb{R}^n$ to the intersection of an affine hyperplane with U.

Fig. 7.3 The intersection of the standard paraboloid in $\mathbb{R}^{n+1}$ with an affine hyperplane projects to a sphere in $\mathbb{R}^n$

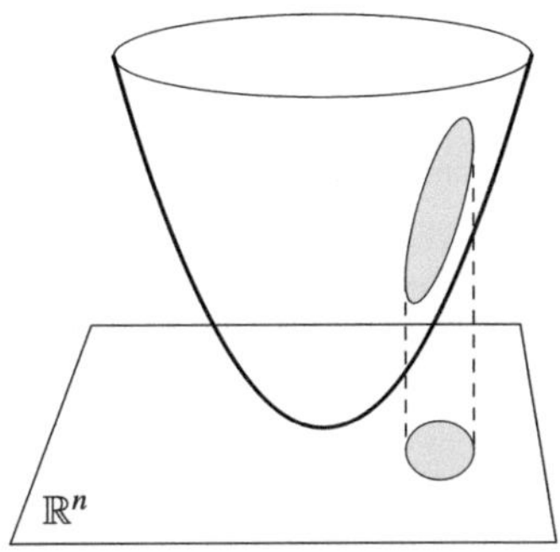

Figure 7.3 illustrates this projection.

Proof For every point $x \in H \cap U$, equating the hyperplane and paraboloid expressions, we obtain the following:

$$2p_1 x_1 + \cdots + 2p_n x_n - p_{n+1} = x_1^2 + \cdots + x_n^2.$$

This implies

$$\sum_{i=1}^{n}(x_i - p_i)^2 = \sum_{i=1}^{n} x_i^2 - 2\sum_{i=1}^{n} x_i\, p_i + \sum_{i=1}^{n} p_i^2 = \sum_{i=1}^{n} p_i^2 - p_{n+1},$$

which is the sphere equation from the statement.

Conversely, every sphere $S \subseteq \mathbb{R}^n$ can be written in the form (7.5), so that the image of S under the lifting $x \mapsto x_U$ is the intersection of U and the hyperplane defined by (7.4). $\qquad\square$

To improve one's understanding of the statement, it may be useful to compare Proposition 7.17 with Lemma 6.11.

Lemma 7.18 *Let $\mathcal{D}$ be a Delone triangulation of S and let $\mathcal{T}$ be a different triangulation of S. Then for all $x \in \operatorname{conv} S$*

$$\psi_{\mathcal{D}}(x) \le \psi_{\mathcal{T}}(x, \Delta)$$

where Δ is an n-simplex from $\mathcal{T}$ that contains x.

Proof Let $\mathbb{S}$ be the sphere spanned by Δ. We can write $\mathbb{S}$ in the form

$$\mathbb{S} = \left\{ x \in \mathbb{R}^n : \sum_{i=1}^{n}(x_i - c_i)^2 = \sum_{i=1}^{n} c_i^2 - c_{n+1} \right\}$$

for a vector $c \in \mathbb{R}^{n+1}$ where $c_{n+1} < \sum_{i=1}^{n} c_i^2$. From this we obtain

$$\psi_{\mathcal{T}}(x, \Delta) = \sum_{i=1}^{n} c_i^2 - c_{n+1} - \sum_{i=1}^{n} (x_i - c_i)^2$$

$$= 2 \sum_{i=1}^{n} c_i x_i - c_{n+1} - \sum_{i=1}^{n} x_i^2. \tag{7.6}$$

The last expression is the directed vertical distance from $x_U = (x_1, \ldots, x_n, \|x\|^2)^T \in U$ to the hyperplane H defined by $x_{n+1} = 2\sum_{i=1}^{n} c_i x_i - c_{n+1}$. Since $\psi_{\mathcal{T}}(x, \Delta) \geq 0$, the hyperplane H lies above x_U, or x_U is a vertex of the Delone polytope $\mathcal{P}^*(S)$. By Proposition 7.17, and since $\mathbb{S}$ contains the $n + 1$ affinely independent points of S,

$$\operatorname{aff}\{x_U : x \in \Delta \cap S\} = H.$$

Thus, the distance (7.6) is minimized if and only if H is a lower supporting hyperplane of $\mathcal{P}^*(S)$. This is equivalent to Δ being a simplex of a Delone triangulation of S. $\qquad\square$

Besides the sphere containing the vertices of an n-simplex Δ, in the following we will study the uniquely determined *smallest enclosing sphere* of Δ. The next exercise illustrates when these two spheres coincide.

Exercise 7.19 The sphere $\mathbb{S}$ spanned by Δ is also the smallest enclosing sphere of Δ if and only if the center of $\mathbb{S}$ is contained in Δ.

We will now show that for an n-simplex Δ, the function $\psi_{\mathcal{T}}(x, \Delta)$ attains its maximum when x is the center of the smallest enclosing sphere of Δ.

Lemma 7.20 *Let $\Delta \in \mathcal{T}$ be an n-simplex with smallest enclosing sphere $\mathbb{S}' = \mathbb{S}(c', \rho')$. Then,*

$$\max_{x \in \Delta} \psi_{\mathcal{T}}(x, \Delta) = \psi_{\mathcal{T}}(c', \Delta) = \rho'^2.$$

Proof Let $\mathbb{S} = \mathbb{S}(c, \rho)$ be the sphere spanned by Δ. If the center c of $\mathbb{S}$ is contained in Δ, by Exercise 7.19 the two spheres $\mathbb{S}$ and $\mathbb{S}'$ coincide, and the statement is clear. Otherwise, c' is contained in the boundary of Δ. Therefore, there exists a unique k-face F of Δ, for $k \in \{0, \ldots, n-1\}$, that contains c in its relative interior. The k-dimensional sphere $\mathbb{S}''$ spanned by F (in $\operatorname{aff} F$) is the intersection of the smallest enclosing sphere $\mathbb{S}'$ and $\operatorname{aff} F$. Here, $\mathbb{S}'$ and $\mathbb{S}''$ have the same center c' (and the same radius ρ'). The point c' minimizes the distance to c, and thus maximizes the function $\psi_{\mathcal{T}}$ on Δ. This is illustrated in Fig. 7.4. It follows that

$$\psi_{\mathcal{T}}(c', \Delta) = \rho^2 - \|c' - c\|^2 = \rho'^2,$$

which proves our claim. $\qquad\square$

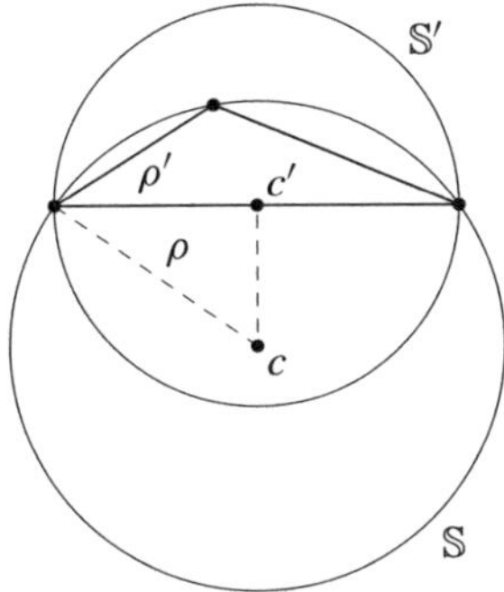

Fig. 7.4 A triangle which spans the circle $\mathbb{S}$ and its smallest enclosing circle $\mathbb{S}'$

Let Δ be a simplex of the triangulation $\mathcal{T}$ of the point set S. We define $\rho(\Delta)$ as the *circumradius*, i.e., the radius of the smallest enclosing sphere of Δ. Then,

$$\rho(\mathcal{T}) := \max_{\Delta \in \mathcal{T}} \rho(\Delta)$$

is the *maximal circumradius* of $\mathcal{T}$.

As mentioned at the beginning of this section, we will show that the Delone triangulations minimize the maximal circumradius in the set of all triangulations of S.

Theorem 7.21 *Let $\mathcal{D}$ be a Delone triangulation of S and let $\mathcal{T}$ be another triangulation of S. Then, $\rho(\mathcal{D}) \leq \rho(\mathcal{T})$.*

Proof Let $x_{\mathcal{T}}$ be a point in conv S that maximizes the function $\psi_{\mathcal{T}}$ and let $x_{\mathcal{D}}$ be a point that maximizes $\psi_{\mathcal{D}}$. By Lemma 7.20, the point $x_{\mathcal{T}}$ is the center of the smallest enclosing sphere $\mathbb{S}(x_{\mathcal{T}}, \rho(\mathcal{T}))$ of an n-simplex Δ in $\mathcal{T}$ which contains $x_{\mathcal{T}}$. In the same way let $\mathbb{S}(x_{\mathcal{D}}, \rho(\mathcal{D}))$ be the smallest enclosing sphere of an n-simplex in $\mathcal{D}$ that contains $x_{\mathcal{D}}$. Using Lemma 7.18 we obtain

$$\rho(\mathcal{D})^2 = \psi_{\mathcal{D}}(x_{\mathcal{D}}) \leq \psi_{\mathcal{T}}\big(x_{\mathcal{D}}, \Delta'\big) \leq \psi_{\mathcal{T}}(x_{\mathcal{T}}, \Delta) = \rho(\mathcal{T})^2,$$

where Δ' is an n-simplex from $\mathcal{T}$ that contains $x_{\mathcal{D}}$. $\square$

Remark 7.22 It is possible for a non-Delone triangulation to have the same maximal circumradius as a Delone triangulation.

7.5 Planar Delone Triangulations

We will again use the strategy of first studying the general case, and then examining the planar case in greater detail. The main result of this section is an algorithm that takes an arbitrary triangulation of a point set $S \subseteq \mathbb{R}^2$ and modifies it step-by-step into a Delone triangulation. This algorithm is not as fast as the beach line algorithm

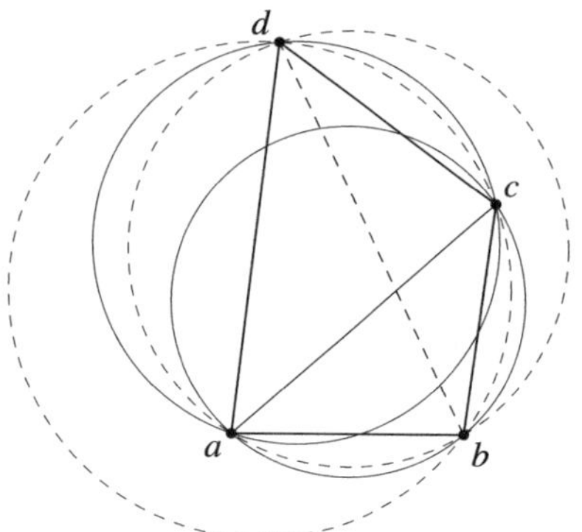

Fig. 7.5 A convex quadrangle with its diagonals and the four circles through each set of three vertices

from Section 6.4, but it is nevertheless interesting for several other reason; see the remarks at the end of this section.

First, we examine an arbitrary planar convex quadrangle with vertices a, b, c, d (in cyclic order). This quadrangle has diagonals $[a, c]$ and $[b, d]$. The four circles through each set of three vertices either coincide, or are pairwise distinct. The latter case occurs when the points are in general position; see Fig. 7.5.

The Delone subdivision of four points in general position is a triangulation. Exactly one of the two diagonals is therefore a Delone edge. By Theorem 7.11, this can be characterized by the existence of a circle through three points from $\{a, b, c, d\}$ that does not contain the fourth point in its interior. The two circles through three points which have the Delone edge as a chord have this property. In Fig. 7.5 the Delone edge is $[a, c]$ and the two *Delone circles* are through a, b, c and a, c, d. The other diagonal and the corresponding *non-Delone circles* are dashed.

The remaining results of this section rely on the following classical result of basic geometry.

Proposition 7.23 (Euclid: *The Elements*, Book III, Proposition 21) *Let $a, b, c, d \in \mathbb{R}^2$ be the vertices of a convex quadrangle in cyclic order. The two diagonals define eight angles $\alpha_1, \alpha_2, \beta_1, \beta_2, \gamma_1, \gamma_2, \delta_1, \delta_2$ as shown in Fig. 7.6. Let C denote the circle through a, b, c. Then d lies*

$$\left\{\begin{array}{l} \text{on the outside of} \\ \text{on} \\ \text{on the inside of} \end{array}\right\} \quad C \quad \text{if and only if} \quad \left\{\begin{array}{l} \alpha_2 > \delta_1 \text{ and } \gamma_1 > \delta_2 \\ \alpha_2 = \delta_1 \text{ and } \gamma_1 = \delta_2 \\ \alpha_2 < \delta_1 \text{ and } \gamma_1 < \delta_2 \end{array}\right\} .$$

Exercise 7.24 In the configuration described in Proposition 7.23 show that the angles $\alpha_2, \beta_1, \beta_2$ and γ_2 are determined by $\alpha_1, \gamma_1, \delta_1$ and δ_2.

An important consequence is that the smallest of the six interior angles of the non-Delone triangulation of a quadrangle is always smaller than the smallest of the six interior angles of a Delone triangulation.

Fig. 7.6 Four points on a circle (*left*; as in Proposition 7.23—congruent angles are identically marked) and a quadrangle with Delone circle (*right*)

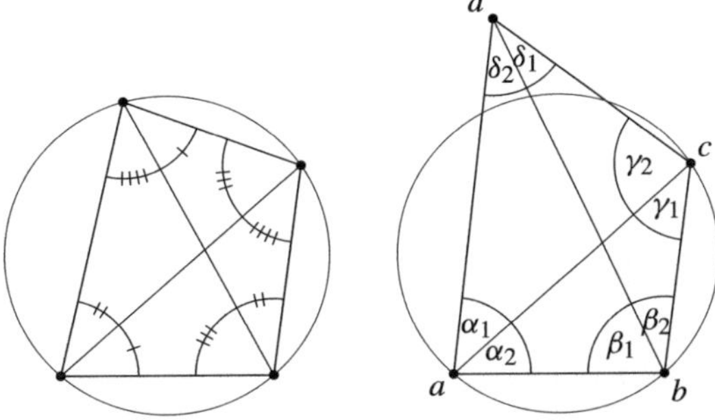

Corollary 7.25 *Let $a, b, c, d \in \mathbb{R}^2$ be the vertices of a convex quadrangle in cyclic order which do not all lie on a common circle. Let $[a, c]$ be the unique Delone edge as in Fig. 7.5. Using the angle labels from Proposition 7.23 and Fig. 7.6, we have*

$$\min\{\alpha_1 + \alpha_2, \beta_1, \beta_2, \gamma_1 + \gamma_2, \delta_1, \delta_2\} < \min\{\alpha_1, \alpha_2, \beta_1 + \beta_2, \gamma_1, \gamma_2, \delta_1 + \delta_2\}. \quad (7.7)$$

Proof We will prove the statement by providing for each element from the second set an element of the first set which is smaller: By Proposition 7.23, $\beta_2 < \alpha_1$, $\delta_1 < \alpha_2$, $\delta_2 < \gamma_1$ and $\beta_1 < \gamma_2$. Since β_2 and δ_2 are positive, $\beta_1 < \beta_1 + \beta_2$ and $\delta_1 < \delta_1 + \delta_2$. $\qquad\qquad\square$

After this examination of the elementary geometry of convex quadrangles, we will now fix a finite point set $S \subseteq \mathbb{R}^2$ that affinely spans the plane, which we will use throughout the remainder of this section.

Let a, b, c, d be points of S such that $\{a, b, c\}$ and $\{a, c, d\}$ are (neighboring) triangles of a triangulation $\mathcal{T}$. If a, b, c, d are the vertices of a convex quadrangle then,

$$\mathrm{Flip}(\mathcal{T}, [a, c]) := \big(\mathcal{T} \setminus \{\mathrm{conv}\{a, b, c\}, \mathrm{conv}\{a, c, d\}, [a, c]\}\big)$$
$$\cup \{\mathrm{conv}\{a, b, d\}, \mathrm{conv}\{b, c, d\}, [b, d]\}$$

is also a triangulation of S. We say $\mathrm{Flip}(\mathcal{T}, [a, c])$ is generated by a *flip* of the edge $[a, c]$ of $\mathcal{T}$. Edge flips are reversible since

$$\mathrm{Flip}\big(\mathrm{Flip}(\mathcal{T}, [a, c]), [b, d]\big) = \mathcal{T}.$$

A *diagonal edge* of a triangulation $\mathcal{T}$ is an edge in $\mathcal{T}$ which is a diagonal in a convex quadrangle consisting of two neighboring triangles in $\mathcal{T}$. We say that the corresponding convex quadrangle is *spanned* by a diagonal edge. A diagonal edge has the *local Delone property* if it is the Delone edge of the quadrangle that it spans. (Such an edge is also said to be *locally Delone*.) The quadrangle which is spanned by a locally Delone diagonal edge satisfies the angle relations from Corollary 7.25, or its vertices lie on a circle (which would imply that the second diagonal is also a Delone edge).

Algorithm 7.2: The flip algorithm for the computation of a Delone triangulation

---

Input: an arbitrary triangulation $\mathcal{T}$ of a finite point set $S \subseteq \mathbb{R}^2$
Output: a triangulation $\mathcal{D}$ of S, such that every diagonal edge has the local Delone property

1 **while** there exists a diagonal edge $e \in \mathcal{T}$ that is not locally Delone **do**
2 $\quad\lfloor\ \mathcal{T} \leftarrow \mathrm{Flip}(\mathcal{T}, e)$

3 **return** $\mathcal{T}$

The usefulness of edge flips for Delone triangulations can be seen in Algorithm 7.2. We will see in Theorem 7.27 that the result is always a Delone triangulation of S.

First, we have to show that Algorithm 7.2 terminates. To do this, we need some sort of quality measure for triangulations of S that increases step-by-step throughout the flip algorithm.

Every triangulation $\mathcal{T}$ of S has the same number of triangles, say k; this will be shown in Exercise 7.29. Therefore, we can assign to $\mathcal{T}$ the vector $W(\mathcal{T})$ of all $3k$ interior angles of $\mathcal{T}$ in increasing order. The lexicographic order of these angle vectors induces a partial order on the set of all triangulations of S. We write $\mathcal{T} > \mathcal{T}'$ if the vector $W(\mathcal{T})$ is larger than $W(\mathcal{T}')$ with respect to the lexicographic order. Since each flip of a diagonal edge which is not locally Delone strictly increases the triangulation, and since there are only a finite number of triangulations of S, the algorithm terminates.

Corollary 7.26 *Let e be a diagonal edge of a triangulation $\mathcal{T}$ of S that is not locally Delone. Then, $\mathrm{Flip}(\mathcal{T}, e) > \mathcal{T}$.*

Proof Under the assumptions of Corollary 7.25, $[b, d]$ is a non-Delone edge of the quadrangle $\mathrm{conv}\{a, b, c, d\}$. The inequality (7.7) states that the non-Delone triangulation $\langle \mathrm{conv}\{a, b, d\}, \mathrm{conv}\{b, c, d\}\rangle$ is smaller than the Delone triangulation

$$\langle \mathrm{conv}\{a, b, c\}, \mathrm{conv}\{a, c, d\}\rangle = \mathrm{Flip}\big(\langle \mathrm{conv}\{a, b, d\}, \mathrm{conv}\{b, c, d\}\rangle, [b, d]\big).$$

This property holds analogously for the quadrangle spanned by e and is inherited by $\mathcal{T}$. All other angles remain constant. $\qquad\square$

We are now able to prove the main theorem of this section which states that the flip algorithm 7.2 computes a Delone triangulation.

Theorem 7.27 *A triangulation $\mathcal{D}$ of S whose diagonal edges satisfy the local Delone property is a Delone triangulation of S.*

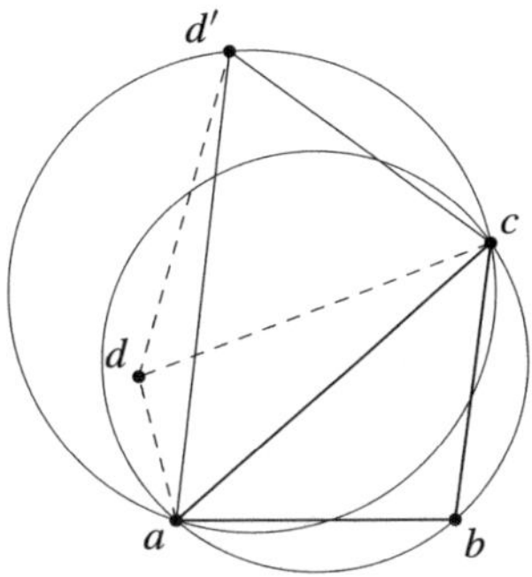

Fig. 7.7 An illustration of
the proof of Theorem 7.27

Proof Assume that the triangulation $\mathcal{D}$ is not a Delone triangulation. Then, by Theorem 7.11, there exists a triangle $\Delta = \mathrm{conv}\{a, b, c\} \in \mathcal{D}$ whose open circumdisk B contains at least one point $d \in S$. Without loss of generality, let $[a, c]$ be the edge of Δ that separates d from Δ. Choose one pair from the set of such pairs (Δ, d) that maximizes the angle (a, d, c). We illustrate this in Fig. 7.7.

The containment $[a, c] \subseteq \mathrm{conv}\{a, b, c, d\}$ implies that $[a, c]$ is a diagonal edge of $\mathcal{D}$ that by assumption satisfies the local Delone property. Thus, there exists a point $d' \in S$ such that $\Delta' := \mathrm{conv}\{a, c, d'\} \in \mathcal{D}$, which lies outside of B. The circumdisk B' of Δ' contains by construction the point d. Also, we have $d \notin \Delta'$, since $\mathcal{D}$ is a triangulation of S. Without loss of generality, let $[a, d']$ be the edge that separates d from Δ'.

Proposition 7.23 implies that the angle (a, d, d') is larger than the angle (a, d, c), which contradicts our choice of the pair (Δ, d) as maximal.   □

In other words, the previous theorem states that a Delone triangulation is a maximal element in the partial order induced by the angle vectors.

Corollary 7.28 *Every Delone triangulation maximizes the smallest interior angle in the set of all triangulations of S.*

In several applications, e.g., finite difference methods for solving partial differential equations, it is desirable to have triangulations including as few narrow triangles as possible. By Corollary 7.28, this, in the planar case, naturally leads to Delone triangulations.

It is possible to show that the Flip Algorithm 7.2 has quadratic worst case runtime. In this sense, it is inferior to the beach line algorithm from Section 6.4. However, the expected run-time (in an appropriate probability model) of the flip algorithm is linear. From a more theoretical viewpoint, the correctness of the algorithm implies that the *configuration space* of all triangulations of a given finite point set is *connected* with respect to flip operations.

Another reason for the flip algorithm's popularity is that it can easily be extended to a *dynamic* algorithm to compute Delone triangulations. By this we mean the fol-

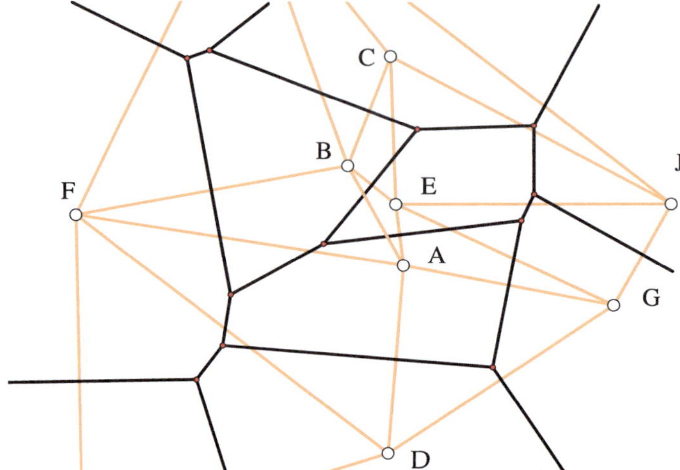

Fig. 7.8 The Voronoi diagram and Delone subdivision of ten points in the plane (the points labeled H and I lie outside of the visible region of $\mathbb{R}^2$)

lowing: Let $S \subseteq \mathbb{R}^2$ and $x \in \mathbb{R}^2 \setminus S$ be such that $S \cup \{x\}$ is in general position. Assume we previously computed a unique Delone triangulation $\mathcal{D}$ of S. Since we assumed $S \cup \{x\}$ to be in general position, x lies in the interior of a triangle $\Delta \in \mathcal{D}$, or on the outside of conv S. In both cases it is easy to modify $\mathcal{D}$ so that we obtain a triangulation of $S \cup \{x\}$. Now, applying the flip algorithm yields a Delone triangulation of $S \cup \{x\}$ after just a few steps.

In a similar way, we can compute a Delone triangulation of $S \setminus \{s\}$ for $s \in S$.

7.6 Inspection Using `polymake`

`polymake` is able to construct Voronoi diagrams and Delone triangulations of arbitrary dimension. We will only deal with the aspects concerning their visualization in this section.

As a first example, we choose the set S to be ten points in the plane whose coordinates represent the locations of the Berlin post offices from the introduction; see Fig. 1.2. To do this, we generate an object `$Postoffices` of type `VoronoiDiagram`. The point set S is given in homogeneous coordinates as the defining property `SITES`. Notice that we prepend the homogenizing ones as a single column vector of length ten. The second property `SITE_LABELS` is optional but useful to identify the points of S in the output.

```
polytope > $S
    = new Matrix([[640,-406],[554,-252],[619,-81],[618,-698],
              [628,-311],[136,-330],[961,-466],[148,-848],
              [392,200],[1049,-308]]);
```

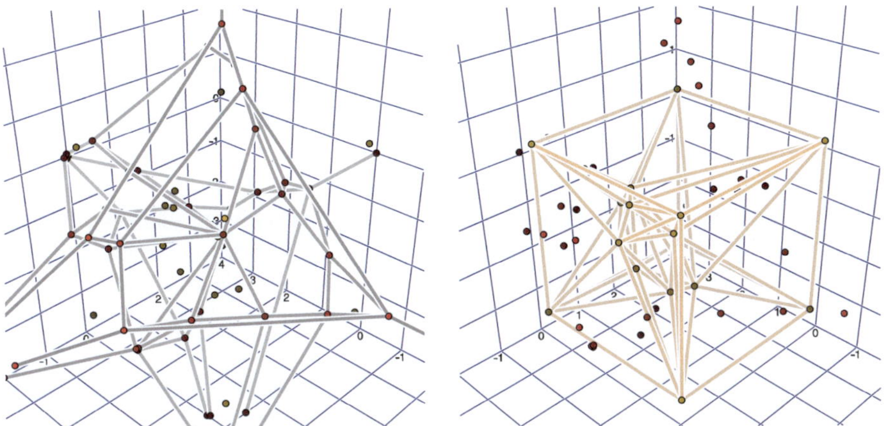

Fig. 7.9 16 points in $\mathbb{R}^3$, their Voronoi diagram (*left*) and corresponding Delone subdivision and Voronoi vertices (*right*). Both pictures only show the region inside the cube $[-4, 4]^3$

```
polytope > $Postoffices
    = new VoronoiDiagram(SITES=>ones_vector(10)|$S,
                         SITE_LABELS=>"A B C D E F G H I J");
```

The command

```
polytope > javaview($Postoffices->VISUAL_VORONOI);
```

initiates the visualization of the Voronoi diagram, and simultaneously the Delone subdivision. Here, by choice, we view the output in `JavaView`, although other output methods are available. The result can be seen in Fig. 7.8. Since we listed specific labels for S in the section `SITE_LABELS`, these labels appear in the output. `polymake` automatically chooses a finite region of $\mathbb{R}^2$ that contains the points of S and all vertices of the Voronoi diagram.

Our second example is 3-dimensional. As a point set we take the eight vertices of a random polytope `$R_3_8` as in Section 3.6.2; see Fig. 3.10, and additionally the eight vertices of the cube with coordinates $\pm 3/2$. Since the vertices of the random polytope are (almost) on the unit sphere, they are contained in the convex hull of the cube's vertices. In total, we have $|S| = 16$.

```
polytope > $R_3_8 = rand_sphere(3,8);
polytope > $C = cube(3,3/2);
polytope > $VD = new VoronoiDiagram(SITES=>($R_3_8->VERTICES/$
                                    C->VERTICES));
polytope > javaview($VD->VISUAL_VORONOI);
```

It is difficult, however, to depict the Voronoi diagram in printed form. The interactive features of `JavaView` are very useful here. Figure 7.9 shows two snapshots which may give the reader an impression of the 3-dimensional image.

7.7 Exercises

Exercise 7.29 Let S be an m-element point set in the plane $\mathbb{R}^2$ such that h points lie on the boundary of the convex hull conv S. Show that every triangulation of S has exactly $2m - 2 - h$ triangles and $3m - 3 - h$ edges.

Exercise 7.30 Show that a triangulation of a finite point set in the plane is a Delone triangulation if and only if for every interior edge e, and for the two triangles which have e as an edge, the sum of the angles which lie opposite e is less than π.

7.8 Remarks

Euclid of Alexandria (ca. 365–300 B.C.) established the axiomatic method in mathematics with his groundbreaking work "The Elements". However, many of the theorems appearing in this work are much older. For example, our Proposition 7.23 is often accredited to Thales of Milet (ca. 624–546 B.C.), indeed it might even be traced back to Babylonian mathematics. We recommend to the reader the interactive version of the "Elements" [68].

Further information about Delone triangulations can be found in [15, 31]. These triangulations were named after the Russian mathematician Boris Nikolajewitsch Delone. Note that several other texts use the name "Delaunay", which comes from a French translation of the name. Delone subdivisions generalize to *regular subdivisions* of polytopes, a concept which is highly relevant to applications in algebraic geometry, for example; see De Loera, Rambau and Santos [32, §2.2.3].

The perturbation procedure of Lemma 3.48 directly gives rise to a triangulation of any polytope. For this there is no need of an additional Delone subdivision as in Exercise 7.14. Triangulations of this kind are known as *pushing triangulations*; see [32, §4.3.4].

Dyer and Frieze showed that computing the volume of a polytope given in outer description is #P-hard [37]. In practical applications it is common to use approximative methods which are based on so-called "random walks"; see Vempala [96] for a good overview.

Part II
Non-linear Computational Geometry

Chapter 8
Algebraic and Geometric Foundations

In the first part of the book we dealt exclusively with polyhedral and hence linear structures. Many situations explicitly or implicitly required computing the intersection of a finite set of affine hyperplanes in the n-dimensional space $\mathbb{R}^n$. This was possible with methods from *linear* algebra. Although it is adequate to use linear geometric structures in many applications, there are also problems which have a natural non-linear representation. We restrict ourselves here to non-linear structures that can be handled with algebraic methods. This chapter is devoted to systems of polynomial equations in two unknowns.

8.1 Motivation

So far we have mainly used the real numbers as a coordinate field. It would seem coherent, therefore, to continue this approach when we begin to look at non-linear geometry. However, it quickly becomes clear that algebraic geometry over the field $\mathbb{R}$ is significantly harder than over its algebraic closure $\mathbb{C}$. Some of our results will be applicable for any field, some will focus exclusively on the complex numbers, and occasionally we will be able to transfer results from the complex to the real numbers.

We begin by studying polynomials in one variable. The roots of a quadratic polynomial $f(x) = x^2 + bx + c$ with real coefficients b, c can be real or complex. Regardless of the root type, we can express them in *terms of radicals*:

$$x_{1,2} = -\frac{b}{2} \pm \sqrt{\frac{b^2}{4} - c}.$$

Similarly, for polynomials f of degree three or four there exist the so-called *Cardano formulas*,[1] which provide explicit expressions for the zeros of f.

---

[1] These formulas are implemented in most computer algebra systems such as `Maple` and `Sage`.

M. Joswig, T. Theobald, *Polyhedral and Algebraic Methods in Computational Geometry*,
Universitext, DOI 10.1007/978-1-4471-4817-3_8,
© Springer-Verlag London 2013

Fig. 8.1 The graph of the function $f(x) = x^5 - 4x + 2$

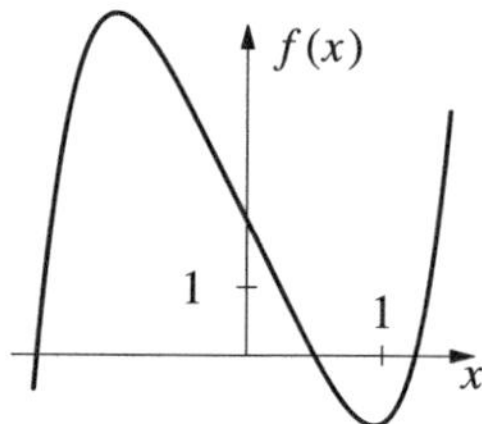

The situation for polynomials of degree ≥ 5, however, is completely different. Galois theory shows that the zeros of a polynomial of degree ≥ 5 are in general not expressible in terms of radicals. An example of such a polynomial (which has the symmetric group of degree 5 as its Galois group) is

$$x^5 - 4x + 2.$$

So we might have to accept that we have to use a polynomial itself to "code" the zeros or to approximate the zeros with numerical methods. In this case we can give an approximation to the complex zeros as

$$-1.518512, \quad 0.508499, \quad 1.243596, \quad -0.116792 \pm 1.438448i.$$

The corresponding real function is illustrated in Fig. 8.1.

Since polynomial systems are an extremely powerful tool in mathematical modeling, it is a central task of geometry to study the sets of zeros of arbitrary polynomials as well as the intersections of those sets. From a computational point of view we focus on computing and manipulating these sets.

Example 8.1 The set of zeros of a quadratic polynomial $f \in \mathbb{R}[x, y]$ defines a conic section (or it is empty). For example, the polynomials

$$f(x, y) = x^2 + y^2 - xy - x - y - 1,$$

$$g(x, y) = 2x^2 - 4y^2 - xy - 2x - 2y - 1$$

define an ellipse and a hyperbola, see Fig. 8.2. It is an important task to characterize the intersection points of such conic sections. Our focus is again on the corresponding computational perspective: In which way(s) can we efficiently and systematically compute these intersection points?

Definition 8.2 For $f \in \mathbb{C}[x_1, \ldots, x_n]$ we call

$$V(f) := \left\{ x \in \mathbb{C}^n : f(x_1, \ldots, x_n) = 0 \right\}$$

the *(complex) affine hypersurface* or *(complex) variety* of f, and we denote by

$$V_{\mathbb{R}}(f) := \left\{ x \in \mathbb{R}^n : f(x_1, \ldots, x_n) = 0 \right\}$$

the *real affine hypersurface* or *real variety* of f.

Fig. 8.2 The intersection of the conic sections defined by f and g (*dashed*)

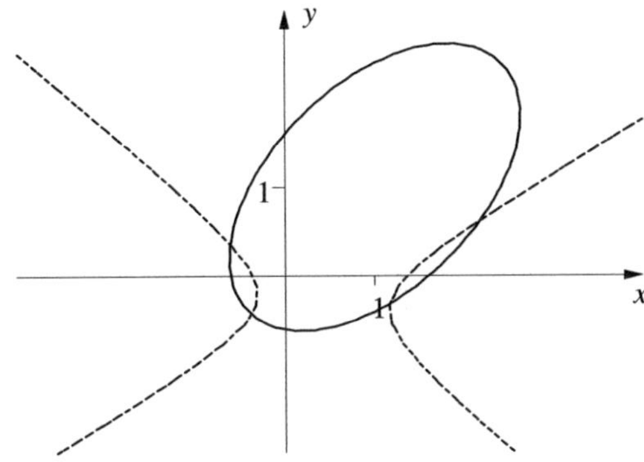

Fig. 8.3 Real hypersurfaces in the plane

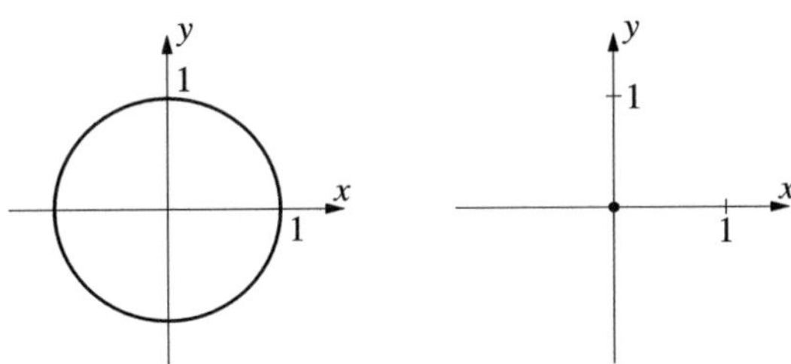

Remark on the notation: In the case of a small number of unknowns we often use $x, y, z, \ldots$ instead of $x_1, x_2, x_3, \ldots$.

Example 8.3 Consider the case $n = 2$. Here we have that

$$
\begin{aligned}
V_{\mathbb{R}}\left(x^2 + y^2 - 1\right) &\quad \text{is a circle;} \\
V_{\mathbb{R}}\left(x^2 + y^2\right) &\quad \text{is a point;} \\
V_{\mathbb{R}}\left(x^2 + y^2 + 1\right) &\quad \text{is empty.}
\end{aligned}
$$

(see Fig. 8.3).

Example 8.4 For $n = 3$ there is a large number of famous examples, including Steiner's Roman Surface

$$
V\left(x^2 y^2 + y^2 z^2 + z^2 x^2 - 2xyz\right) \tag{8.1}
$$

and Clebsch's Diagonal Surface

$$
V\left(16x^3 + 16y^3 - 31z^3 + 24x^2 z - 48x^2 y - 48xy^2 + 24y^2 z - 54\sqrt{3}z^2 - 72z\right). \tag{8.2}
$$

The real parts of these surfaces are depicted in Fig. 8.4.

Contrary to the intuition we get from the illustration, the algebraic surface defined by the polynomial (8.1) contains the three coordinate axes as singular loci. We can see this by directly examining the equation.

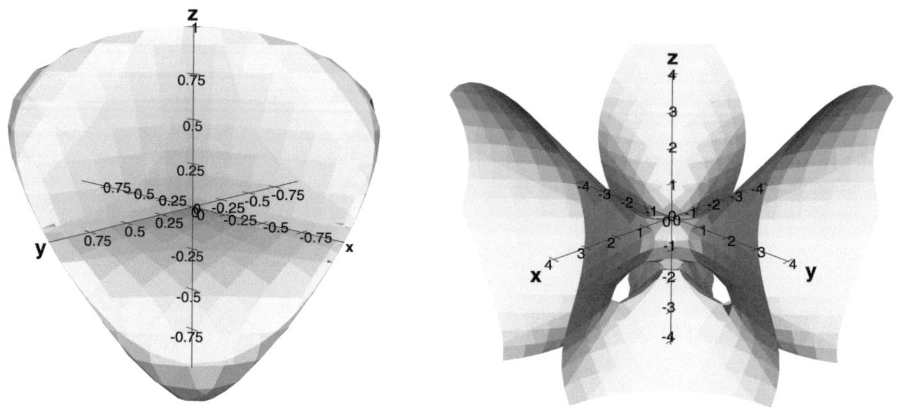

Fig. 8.4 *Left*: Steiner's Roman Surface (8.1), *right*: Clebsch's Diagonal Surface (8.2)

8.2 Univariate Polynomials

We will again study the case of polynomials in one unknown. As mentioned before, the roots of univariate polynomials of degree ≥ 5 are in general not expressible in terms of radicals. For the numerical approximation of roots it is necessary to distinguish between the task of computing just one and computing all of the roots of a polynomial. Also, it may happen (e.g. in the case of coefficients which vary extremely in size) that numerical methods are badly conditioned and will not converge.

It is possible to formulate the computation of all roots of a univariate polynomial over an arbitrary field K as an eigenvalue problem of linear algebra. For the computation of the eigenvalues of a complex matrix numerous well-studied numerical methods are available.

The eigenvalues of a matrix $A \in K^{n \times n}$ are the roots of the *characteristic polynomial* of A, i.e., the roots of

$$\chi_A(t) = \det(A - tI),$$

where $I \in K^{n \times n}$ is the identity matrix. The characteristic polynomial $p(t)$ is always of degree n with leading coefficient $(-1)^n$. In order to formulate the computation of the roots of an arbitrary polynomial p as an eigenvalue problem, it is sufficient to find a matrix A with characteristic polynomial p.

Definition 8.5 The *companion matrix* of the normalized polynomial

$$p(t) = t^n + a_{n-1}t^{n-1} + \cdots + a_1 t + a_0 \in K[t]$$

of degree n is the matrix

$$C_p = \begin{pmatrix} 0 & 1 & 0 & \cdots & 0 \\ 0 & 0 & 1 & \cdots & 0 \\ \vdots & \vdots & \vdots & \ddots & \vdots \\ 0 & 0 & 0 & \cdots & 1 \\ -a_0 & -a_1 & -a_2 & \cdots & -a_{n-1} \end{pmatrix} \in K^{n \times n}.$$

Theorem 8.6 *The characteristic polynomial of the companion matrix of the normalized polynomial*

$$p(t) = t^n + a_{n-1}t^{n-1} + \cdots + a_1 t + a_0 \in K[t]$$

of degree $n \geq 1$ is

$$\det(C_p - tI) = (-1)^n p(t).$$

Proof The proof is by induction on n. For $n = 1$ the statement is obvious and for $n > 1$, eliminating the first row and the first column of C_p gives the companion matrix of the polynomial $q(t) = t^{n-1} + a_{n-1}t^{n-2} + \cdots + a_2 t + a_1$. Hence we can write

$$\det(C_p - tI) = (-t)(-1)^{n-1}q(t) + (-1)^{n+1}(-a_0),$$

which leads to

$$\det(C_p - tI) = (-1)^n p(t). \qquad \square$$

8.3 Resultants

Let K be an arbitrary field. Using the resultant of two polynomials $f, g \in K[x]$, we can decide if f and g have a common factor of positive degree without explicitly computing this factor. If K is algebraically closed, the existence of a non-trivial common factor is equivalent to f and g having a common zero.

Definition 8.7 Let $n, m \geq 1$ and

$$f = a_n x^n + \cdots + a_1 x + a_0 \quad \text{and}$$
$$g = b_m x^m + \cdots + b_1 x + b_0$$

be polynomials of degree n and m in $K[x]$. The *resultant* $\mathrm{Res}(f, g)$ is the determinant of the $(m + n) \times (m + n)$-matrix

$$
\left.
\begin{pmatrix}
a_n & a_{n-1} & \cdots & a_0 & & & \\
 & \ddots & \ddots & & & \ddots & \\
 & & a_n & a_{n-1} & \cdots & a_0 \\
b_m & b_{m-1} & \cdots & b_0 & & & \\
 & \ddots & \ddots & & & \ddots & \\
 & & b_m & b_{m-1} & \cdots & b_0
\end{pmatrix}
\right\}
\begin{matrix}
\\ m \text{ rows,} \\ \\ \\ n \text{ rows.}
\end{matrix}
\tag{8.3}
$$

The matrix (8.3) is called the *Sylvester matrix* of f and g.

Theorem 8.8 *Two polynomials $f, g \in K[x] \setminus \{0\}$ of positive degrees have a common factor of positive degree if and only if $\mathrm{Res}(f, g) = 0$.*

To prove this, we first need to show the following:

Lemma 8.9 *The resultant $\mathrm{Res}(f, g)$ of two polynomials $f, g \in K[x]$ with positive degrees vanishes if and only if there exist polynomials $r, s \in K[x]$ with $(r, s) \neq (0, 0)$, $\deg r < \deg f$, $\deg s < \deg g$ and $sf + rg = 0$.*

Proof We interpret the rows of the Sylvester matrix as vectors

$$
x^{m-1} f, \ldots, xf, f, x^{n-1} g, \ldots, xg, g
$$

in the K-vector space of polynomials of degree $< m + n$ (with respect to the basis $x^{m+n-1}, x^{m+n-2}, \ldots, x, 1$). The resultant $\mathrm{Res}(f, g)$ vanishes if and only if these $m + n$ vectors are linearly dependent, i.e., if there exist coefficients $r_0, \ldots, r_{n-1}$ and $s_0, \ldots, s_{m-1}$ in K which do not simultaneously vanish and such that

$$
s_{m-1} x^{m-1} f + \cdots + s_1 xf + s_0 f + r_{n-1} x^{n-1} g + \cdots + r_1 xg + r_0 g = 0.
$$

Using the notation $r := \sum_{i=0}^{n-1} r_i x^i$ and $s := \sum_{j=0}^{m-1} s_j x^j$ this is the case if and only if $(r, s) \neq (0, 0)$, $\deg r < \deg f$, $\deg s < \deg g$ and $sf + rg = 0$. $\qquad\square$

Proof of Theorem 8.8 We show that f and g have a non-constant common factor if and only if they satisfy the condition from Lemma 8.9. If f and g have a common non-constant factor $h \in K[x]$ then there exist polynomials $f_0, g_0 \in K[x]$ with

$$
f = hf_0 \quad \text{and} \quad g = hg_0,
$$

and we can choose $r := f_0$ and $s := -g_0$.

To prove the reverse implication it is important to note that every non-constant polynomial in $K[x]$ can be uniquely written as a product of prime factors. From the prime factor decomposition of the four polynomials in the equation $sf = -rg$ we obtain the equation

$$s_1 \cdots s_k \cdot f_1 \cdots f_p = -r_1 \cdots r_l \cdot g_1 \cdots g_q, \tag{8.4}$$

which may contain constant factors as well. Without loss of generality we assume that s_1, f_1, r_1, g_1 are constant and that all other prime factors are normalized polynomials of positive degree. Further, we can assume that $s \neq 0$ (otherwise we switch the roles of f and g and of r and s). Therefore we have that $\deg g > \deg s \geq 0$, and hence $q \geq 2$ and g_2 is a normalized prime factor of g with positive degree. Since $\deg g > \deg s$ and since the prime factor decomposition is unique, there exists at least one normalized prime factor g_j of g that also appears in $f_2, \ldots, f_p$. Hence g_j is a non-constant common factor of f and g. $\qquad\square$

As mentioned above, the proof relies on the fact that the polynomial ring $K[x]$ is a *unique factorization domain*. That is, $K[x]$ is commutative, does not contain zero divisors and every polynomial in $K[x]$ has a unique decomposition into prime factors; see Appendix A. Gauss' Lemma, Theorem A.4, shows that for every unique factorization domain R, the polynomial ring $R[x]$ is also a unique factorization domain.

By analyzing the proofs in this section, one can see that all of the results hold for a polynomial ring $R[x]$ over an arbitrary unique factorization domain R. Note that R, being zero divisor free and commutative, has a quotient field that we denote by K. The linear algebraic methods which we used can then be interpreted with respect to this field. Note furthermore that the polynomials r and s in Lemma 8.9 can be chosen in $R[x]$ for any $f, g \in R[x]$, since we could otherwise simply multiply the equation $sf + rg = 0$ by the lowest common denominator of r and s.

These abstract remarks are relevant since they imply that the statements of this section also hold for polynomial rings $K[x_1, \ldots, x_n]$ in several variables over a field K. To see this, note that

$$K[x_1, \ldots, x_n] = \big(K[x_1, \ldots, x_{n-1}]\big)[x_n]. \tag{8.5}$$

When writing the resultant of two multivariate polynomials we have to keep track of the variable which we use to build the resultant. In the case of (8.5) we write, for example, Res_{x_n} and analogously $\deg_{x_n}$ for the degree in the unknown x_n.

We summarize this result with a corollary.

Corollary 8.10 *Two polynomials $f, g \in K[x_1, \ldots, x_n]$ of positive degree in x_n have a common factor of positive degree in x_n if and only if $\mathrm{Res}_{x_n}(f, g)$ is the zero polynomial in $K[x_1, \ldots, x_{n-1}]$.*

8.4 Plane Affine Algebraic Curves

Some of the simplest examples of non-linear structures are algebraic curves in the plane. We will now examine the complex case.

Definition 8.11 A subset $C \subseteq \mathbb{C}^2$ is called an *affine-algebraic curve* if there exists a non-constant polynomial $f \in \mathbb{C}[x, y]$ such that

$$C = V(f) = \left\{ (x, y) \in \mathbb{C}^2 : f(x, y) = 0 \right\}.$$

Obviously the polynomial f for a given hypersurface is not determined uniquely, since we have for any $\lambda \in \mathbb{C} \setminus \{0\}$ and $k \geq 1$ that $V(f) = V(\lambda f) = V(f^k)$. An important result of this section will be that this is the only type of uncertainty that occurs. However, the situation is completely different when we restrict ourselves to the real numbers. One of the easiest ways to see this is by observing that the empty set may be written in various ways as a real hypersurface.

Every non-constant polynomial $f \in \mathbb{C}[x, y]$ defines a curve $V(f) \subseteq \mathbb{C}^2$. If f is a divisor of g, i.e., $g = f \cdot h$ for a polynomial h, then $V(f) \subset V(g)$ and $V(g) = V(f) \cup V(h)$. The following Lemma by Study, a predecessor of Hilbert's Nullstellensatz (see Section 10.4), enables us to use the point sets of curves to gain insight about the divisibility of polynomials.

Lemma 8.12 (Study's Lemma) *Let $f, g \in \mathbb{C}[x, y]$. If f is irreducible, not constant, and $V(f) \subseteq V(g)$, then f divides g.*

Proof Let

$$f = a_n y^n + \cdots + a_1 y + a_0,$$

$$g = b_m y^m + \cdots + b_1 y + b_0$$

be polynomials in $\mathbb{C}[x, y]$ with coefficients $a_i, b_j \in \mathbb{C}[x]$. If $f, g \in K[x]$, i.e., $f = a_0$ and $g = b_0$, the statement is true. So we can assume (possibly after switching x and y) that $n \geq 1$. We claim that $m \geq 1$. If not, then there exists an $\alpha \in \mathbb{C}$ with $a_n(\alpha) \neq 0$ and $b_0(\alpha) \neq 0$, since the univariate polynomials a_n and b_0 have only finitely many zeros. But this implies that $V(f)$ and the line $[-\alpha : 1 : 0]$ intersect. This in turn implies that $V(g) \cap [-\alpha : 1 : 0] = \emptyset$ and this contradicts our assumption that $V(f) \subseteq V(g)$. We illustrate this in Fig. 8.5.

We will now show that the resultant $\mathrm{Res}_y(f, g)$ has infinitely many zeros $x \in \mathbb{C}$. Since $\mathrm{Res}_y(f, g)$ is a polynomial in x this implies that it is the zero polynomial and hence that f and g have a common divisor. As f was assumed to be irreducible, it follows that f divides g.

In the following we study only those $\alpha \in \mathbb{C}$ with $a_n(\alpha) \neq 0$ and $b_m(\alpha) \neq 0$. Doing this we exclude finitely many $\alpha \in \mathbb{C}$ since $a_n \neq 0$ and $b_m \neq 0$. Plugging $x = \alpha$ in f and g results in polynomials $f_\alpha, g_\alpha \in \mathbb{C}[y]$. If f_α has pairwise distinct zeros $c_1, \ldots, c_k \in \mathbb{C}$, then they are also zeros of g_α. Therefore, since $1 \leq k \leq n$,

$$(y - c_1) \cdots (y - c_k)$$

is a non-constant common factor of f_α and g_α in $\mathbb{C}[y]$. It follows that

$$\big(\mathrm{Res}_y(f, g) \big)(\alpha) = \mathrm{Res}_y(f_\alpha, g_\alpha) = 0. \qquad \square$$

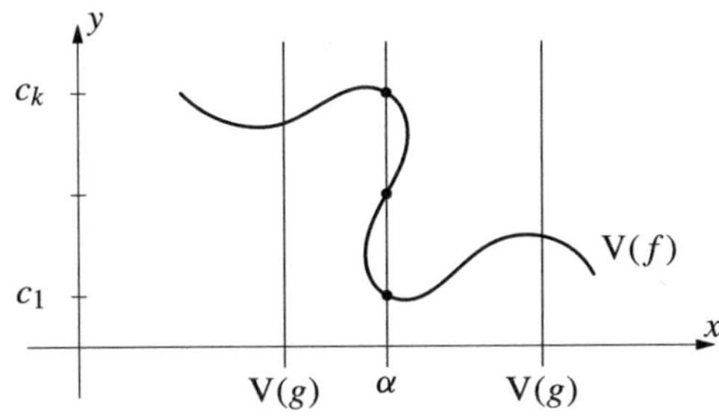

Fig. 8.5 Illustration under the assumption that $m = 0$

8.5 Projective Curves

When studying algebraic curves (and algebraic hypersurfaces in general) it is useful to view them as objects in projective space. A reason for this is that the point set of the projective space $\mathbb{P}^n_{\mathbb{C}}$ is compact, while that of $\mathbb{C}^n$ is not.

Example 8.13 The complex standard parabola $V(x^2 - y)$ intersects a given line L in at most two points. If the line is given in the form $V(ax + b - y)$ with $a, b \in \mathbb{C}$ then we can compute the intersection points by comparing the equations of the parabola and the line.

The degenerate case appears when the line L lies tangent to $V(x^2 - y)$. In this case we have a double intersection point (in the sense of the definition in Section 8.6). So, counting multiplicity, we still have two intersection points.

If $L = V(x - c)$ for a constant $c \in \mathbb{C}$ is a vertical line, the parabola and the line intersect in only one point (see Fig. 8.6). Since the line is not tangent to the parabola, this intersection point has multiplicity 1. In the following we will see that in our example there is a second intersection point which is "invisible" in the affine plane $\mathbb{C}^2$. This point lies on the ideal line of the projective plane $\mathbb{P}^2_{\mathbb{C}}$.

Note that the illustration of the situation in real space, as in Fig. 8.6, can be misleading: The possibility of the line not intersecting the parabola does not occur in complex space.

Curves in the projective plane are defined by homogeneous polynomials $f \in \mathbb{C}[w, x, y]$.

Definition 8.14

(a) A polynomial $f \in \mathbb{C}[x_1, \ldots, x_n]$ is *homogeneous* of degree d if for all $\lambda \in \mathbb{C}$ we have

$$f(\lambda x_1, \ldots, \lambda x_n) = \lambda^d f(x_1, \ldots, x_n).$$

(b) Let $f \in \mathbb{C}[x_1, \ldots, x_n]$ be a homogeneous polynomial. Then

$$V(f) := \left\{ (a_1 : \cdots : a_n)^T \in \mathbb{P}^{n-1}_{\mathbb{C}} : f(a_1, \ldots, a_n) = 0 \right\}$$

is called the *(complex) projective hypersurface* of f.

Fig. 8.6 The intersection of a
parabola with a *vertical line*

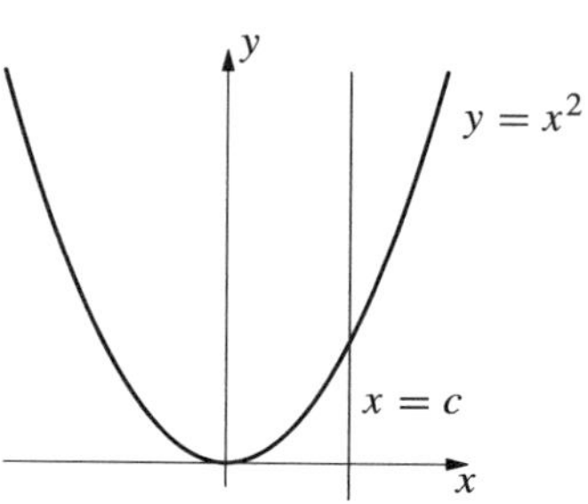

(c) The *total degree* of a monomial $x_1^{\alpha_1} \cdots x_n^{\alpha_n}$ is the sum of its exponents $\alpha_1 + \cdots + \alpha_n$. The *total degree* tdeg f of a polynomial f is the maximum of the total degrees of its monomials (where we set tdeg $0 = -\infty$).

(d) The *degree* of a projective hypersurface $V \subseteq \mathbb{P}^n_{\mathbb{C}}$ is the maximum of the total degrees of all homogeneous $f \in \mathbb{C}[x_0, x_1, \ldots, x_n]$ with $V(f) = V$.

Note that the condition $f(x_1, \ldots, x_n) = 0$ in Definition 8.14b. is independent of the choice of the homogeneous coordinates $(a_1 : \cdots : a_n)$ since

$$f(\lambda a_1, \ldots, \lambda a_n) = 0 \quad \Longleftrightarrow \quad f(a_1, \ldots, a_n) = 0$$

for all $\lambda \in \mathbb{C} \setminus \{0\}$ due to the homogeneity of f.

A *projective algebraic curve* is the projective hypersurface of a non-constant homogeneous polynomial in $\mathbb{C}[w, x, y]$ without repeated factors. We can directly see that a polynomial is homogeneous of degree d if each of its monomials has total degree d.

Remark 8.15 For a not necessarily homogeneous polynomial f we define the *homogeneous component of degree d* as the sum of all terms of total degree d. Every polynomial is the sum of its homogeneous components.

We will now illustrate the usefulness of the projective approach. To do this we write a polynomial $f \in \mathbb{C}[x_1, \ldots, x_n]$ as the sum of its monomials

$$f = \sum_{i \in I} c_i x_1^{\alpha_1^{(i)}} \cdots x_n^{\alpha_n^{(i)}}. \tag{8.6}$$

Here the $c_i \in \mathbb{C} \setminus \{0\}$ are the coefficients and I is a (finite) set that serves as an index set for the monomials. We abbreviate the total degree of the monomial $x_1^{\alpha_1^{(i)}} \cdots x_n^{\alpha_n^{(i)}}$ as $d_i := \alpha_1^{(i)} + \cdots + \alpha_n^{(i)}$. Then

$$d := \text{tdeg } f = \max\{d_i : i \in I\}$$

is the total degree of f. The polynomial

$$\bar{f} := \sum_{i \in I} c_i x_0^{d-d_i} x_1^{\alpha_1^{(i)}} \cdots x_n^{\alpha_n^{(i)}} \in \mathbb{C}[x_0, x_1, \ldots, x_n] \tag{8.7}$$

is homogeneous of degree d and is called the *homogenization* of f.

In (2.1) we defined the map $\iota : \mathbb{C}^n \to \mathbb{P}_{\mathbb{C}}^n$, $(x_1, \ldots, x_n)^T \mapsto (1 : x_1 : \cdots : x_n)^T$, which embeds the affine space $\mathbb{C}^n$ in its projective closure $\mathbb{P}_{\mathbb{C}}^n$. Using ι we can treat affine hypersurfaces as subsets of the projective space.

Proposition 8.16 *Let $f \in \mathbb{C}[x_1, \ldots, x_n]$ be an arbitrary non-constant polynomial with homogenization $\bar{f} \in \mathbb{C}[x_0, x_1, \ldots, x_n]$. Then we have*

$$\iota\big(\mathrm{V}(f)\big) = \mathrm{V}(\bar{f}) \cap \iota\big(\mathbb{C}^n\big).$$

Proof For $a \in \mathbb{C}^n$ we have $\iota(a) = (1 : a_1 : \cdots : a_n)$. Using the notation of (8.6) and (8.7) we have that

$$f(a) = \sum_{i \in I} c_i a_1^{\alpha_1^{(i)}} \cdots a_n^{\alpha_n^{(i)}} = \sum_{i \in I} c_i 1^{d-d_i} a_1^{\alpha_1^{(i)}} \cdots a_n^{\alpha_n^{(i)}} = \bar{f}\big(\iota(a)\big).$$

In particular, $f(a)$ vanishes if and only if $\bar{f}(\iota(a))$ vanishes. $\square$

We call $\mathrm{V}(\bar{f})$ the *projective closure* of $\mathrm{V}(f)$.

Example 8.17 Recall Example 8.13. Let $f = x^2 - y \in \mathbb{C}[x, y]$ and $\bar{f} = x^2 - wy \in \mathbb{C}[w, x, y]$ be its homogenization. Respectively we have that $x - cw$ is the homogenization of the linear polynomial $x - c$ for $c \in \mathbb{C}$. The projective closure of the affine line $L = \mathrm{V}(x - c)$ is the projective line $[-c : 1 : 0] = \mathrm{V}(x - cw)$. Its point at infinity $(0 : 0 : 1)^T$ is contained in $\mathrm{V}(x^2 - wy)$ since $0^2 - 0 \cdot 1 = 0$. This is the "missing" intersection point we were looking for.

Exercise 8.18 Show that for every projective transformation $\pi : \mathbb{P}_{\mathbb{C}}^n \to \mathbb{P}_{\mathbb{C}}^n$ and every non-constant polynomial $f \in \mathbb{C}[x_0, x_1, \ldots, x_n]$ the image $\pi(\mathrm{V}(f))$ of the projective hypersurface f under π is again a projective hypersurface of the same degree.

8.6 Bézout's Theorem

In this section we study how two projective (algebraic) curves C and D of degrees n and m intersect in the complex projective plane. We will show that C and D intersect in at most nm points unless they have a common component.

Definition 8.19

(a) The curve C is said to be *irreducible* if there exists an irreducible polynomial that defines C.
(b) Let $f, g \in \mathbb{C}[w, x, y]$ be homogeneous. If g divides f we call $\mathrm{V}(g)$ a *component* of $\mathrm{V}(f)$.

First, we will study the special case of a curve intersecting a line as in Example 8.13. Let $C = \mathrm{V}(f) \subseteq \mathbb{P}^2_{\mathbb{C}}$ be a curve with homogeneous polynomial $f \in \mathbb{C}[w, x, y]$ of degree n. To simplify the computations we assume that the line L is given by $\mathrm{V}(y) = \{(a : b : 0)^T \in \mathbb{P}^2_{\mathbb{C}} : (a : b)^T \in \mathbb{P}^1_{\mathbb{C}}\}$. The intersection points of C and L satisfy

$$C \cap L = \{(a : b : 0)^T \in \mathbb{P}^2_{\mathbb{C}} : f(a, b, 0) = 0\}.$$

As in (8.5), f can be expressed as

$$f(w, x, y) = f_n y^n + f_{n-1} y^{n-1} + \cdots + f_0$$

with coefficients $f_0, \ldots, f_n \in \mathbb{C}[w, x]$ and $\mathrm{tdeg}\, f_i = n - i$ (or $f_i = 0$). This gives $f(w, x, 0) = f_0(w, x)$.

We distinguish between two cases: For $f_0 = 0$, y divides f and $L \subseteq C$ follows. If $f_0 \neq 0$ we have that $\deg f_0 = n$ since f is homogeneous of degree n. By the fundamental theorem of algebra (in its homogeneous formulation) there exists a decomposition (unique up to the order)

$$f_0 = (b_1 w - a_1 x)^{k_1} \cdots (b_m w - a_m x)^{k_m}$$

with uniquely determined (up to their order) pairwise distinct points $(a_i : b_i)^T \in \mathbb{P}^1_{\mathbb{C}}$ and $k_i \in \mathbb{N}$ for $1 \leq i \leq m$. We define

$$\mathrm{mult}_p(C, L) := \begin{cases} k_i & \text{for } p = (a_i : b_i : 0)^T \text{ for an } i \in \{1, \ldots, m\}, \\ 0 & \text{for } p \notin \{(a_i : b_i : 0)^T : 1 \leq i \leq m\} \end{cases}$$

as the *intersection multiplicity* of C and L in the point p.

Remark 8.20 We showed in Exercise 8.18 that projective transformations map projective hyperplanes to projective hyperplanes of the same degree. Since every projective line can be transformed to any other projective line, we see that the above definition of multiplicity is valid for the intersection of a curve with an arbitrary line.

Lemma 8.21 *Let $C \subseteq \mathbb{P}^2_{\mathbb{C}}$ be a curve of degree $n \geq 1$ and L a line which is not contained in C. Then the number of intersection points of C and L counting multiplicity is equal to n.*

Proof This follows immediately from $k_1 + \cdots + k_m = n$. $\square$

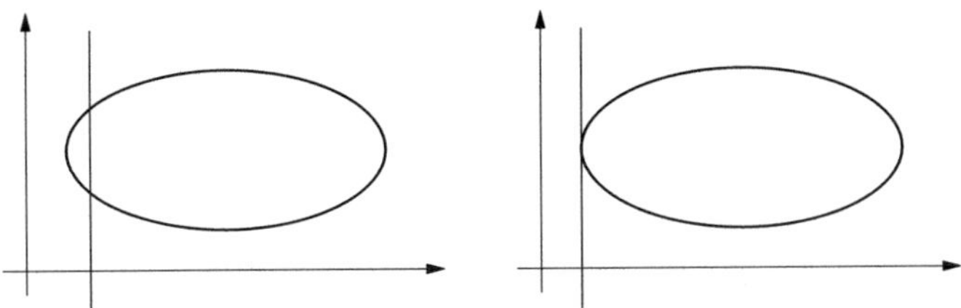

Fig. 8.7 Intersection points of multiplicity 1 and 2

Example 8.22 The picture in Fig. 8.7 shows two intersection points of multiplicity 1 and the right picture shows an intersection point of multiplicity 2.

Exercise 8.23 Let $f \in \mathbb{C}[x]$ be a non-constant polynomial of degree n. Show that the algebraic curve $V(f(x) - y) \subseteq \mathbb{C}^2$ and the line $V(y)$ have a $(k+1)$-fold intersection point at $(\alpha, 0)$ if and only if α is a *zero of order* $(k+1)$, i.e., α is a zero of f and of all derivatives $f^{(1)} = f'$, $f^{(2)} = f''$, $\ldots$, $f^{(k)} = f^{(k-1)'}$. In particular we have that the sum of all orders of all zeros equals n.

We further clarify the multiplicity of intersection points of arbitrary curves below. First, we will prove a weaker form of Bézout's theorem.

Theorem 8.24 (Weak form of Bézout's Theorem) *If two projective curves $C, D \subseteq \mathbb{P}^2_{\mathbb{C}}$ of degrees n and m do not have a common component, then they intersect in at most nm points.*

To prove this we need the following technical lemma.

Lemma 8.25 *Let $f, g \in \mathbb{C}[w, x, y]$ be non-constant homogeneous polynomials of degrees n and m with*

$$f(0, 0, 1) \neq 0 \neq g(0, 0, 1). \tag{8.8}$$

Then f and g have a non-constant common factor if and only if the resultant $\mathrm{Res}_y(f, g) \in \mathbb{C}[w, x]$ is the zero polynomial. If f and g have no non-constant common factor then $\mathrm{Res}_y(f, g)$ has degree nm.

Exercise 8.26 Show that the technical assumption $f(0, 0, 1) \neq 0$ guarantees that the degree of the homogeneous polynomial $f \in \mathbb{C}[w, x, y]$ equals the degree of f interpreted as a polynomial in y with coefficients in $\mathbb{C}[w, x]$.

Again, by Exercise 8.18 the assumption (8.8) is irrelevant.

Proof of Lemma 8.25 The first statement follows from the homogeneous version of Corollary 8.10. For the second statement we first remark that the resultant

$\text{Res}_y(f, g)$ is the determinant of an $(n + m) \times (n + m)$-matrix whose non-zero entries r_{ij} in row i and column j are homogeneous polynomials in $\mathbb{C}[w, x]$ of degree d_{ij} with

$$d_{ij} = \begin{cases} j - i & \text{if } 1 \leq i \leq m, \\ j - i + m & \text{if } m + 1 \leq i \leq n + m. \end{cases}$$

Then $\text{Res}_y(f, g)$ is a sum of terms of the form

$$\pm \prod_{i=1}^{n+m} r_{i,\sigma(i)},$$

where σ is a permutation of $\{1, \ldots, n + m\}$. Each of these terms is either the zero polynomial or a homogeneous polynomial of degree

$$\sum_{i=1}^{m+n} d_{i,\sigma(i)} = \sum_{i=1}^{m} (\sigma(i) - i) + \sum_{i=m+1}^{n+m} (\sigma(i) - i + m)$$

$$= nm - \sum_{i=1}^{n+m} i + \sum_{i=1}^{n+m} \sigma(i)$$

$$= nm.$$

Since the resultant is not the zero polynomial it must be of degree nm. $\qquad\square$

Proof of Theorem 8.24 As in the proof of Study's lemma we sweep over the plane with a set of lines. Without loss of generality we can assume that the curves C and D do not contain the point $(0 : 0 : 1)^T$. Consider the representation

$$f = a_n y^n + a_{n-1} y^{n-1} + \cdots + a_0,$$

$$g = b_m y^m + b_{m-1} y^{m-1} + \cdots + b_0$$

with coefficients $a_i, b_j \in \mathbb{C}[w, x]$. Since f and g are homogeneous of degrees m and n we have that $\deg a_i = n - i$ and $\deg b_j = m - j$ if $a_i, b_j \neq 0$. $(0 : 0 : 1)^T \notin C \cup D$ implies $a_n \neq 0$ and $b_m \neq 0$. As, by assumption, C and D do not have a common component, Lemma 8.25 tells us that $r := \text{Res}_y(f, g)$ is a homogeneous polynomial of degree nm in $\mathbb{C}[w, x]$.

To show that $C \cap D$ is finite we have to study the resultant in greater detail. Substituting an arbitrary fixed point $(\alpha : \beta)^T$ of the projective line $\mathbb{P}^1_{\mathbb{C}}$ into f and g for x and w respectively, we obtain polynomials $f_{(\alpha:\beta)}$ and $g_{(\alpha:\beta)}$ in $\mathbb{C}[y]$. We now apply Theorem 8.8 to these two univariate polynomials.

A point $(\alpha : \beta)^T \in \mathbb{P}^1_{\mathbb{C}}$ is a zero of the resultant r if and only if there exists a $\gamma \in \mathbb{C}$ such that $(\alpha : \beta : \gamma)^T \in V(f) \cap V(g)$. Since we already know that $r \neq 0$, we have that r can only have finitely many zeros on the projective line. For any fixed zero $(\alpha : \beta)^T$ of r there exist only finitely many γ with $f(\alpha, \beta, \gamma) = g(\alpha, \beta, \gamma) = 0$. Otherwise the line

$$[-\beta : \alpha : 0] = \left\{\lambda(\alpha : \beta : 0)^T + \mu(0 : 0 : 1)^T : (\lambda : \mu)^T \in \mathbb{P}^1_{\mathbb{C}}\right\}$$
$$= \left\{(\lambda\alpha : \lambda\beta : \mu)^T : (\lambda : \mu)^T \in \mathbb{P}^1_{\mathbb{C}}\right\}$$

connecting $(\alpha : \beta : 0)^T$ and $(0 : 0 : 1)^T$ would be a common component of C and D. Hence $C \cap D$ is finite.

Between the finitely many intersection points we can have only finitely many connecting lines. Using a proper projective transformation we can guarantee that the point $(0 : 0 : 1)^T$ is not contained in any of the connecting lines. So each line $[-\beta : \alpha : 0]$ contains at most one intersection point of C and D and there are at most $nm = \deg r$ intersection points. $\square$

For a stronger statement we need to define the multiplicity of an intersection point of two arbitrary curves $C = V(f)$ and $D = V(g)$ which have no common component. As at the end of the proof of Theorem 8.24, we assume that each line $[-\beta : \alpha : 0]$ with $(\alpha : \beta)^T \in \mathbb{P}^1_{\mathbb{C}}$ contains at most one point from $C \cap D$ and that $(0 : 0 : 1)^T \notin C \cup D$.

The key concept for the following statements is the resultant $r = r_y(f, g)$, whose zeros on the projective line $\mathbb{P}^1_{\mathbb{C}} = [0 : 0 : 1] \subseteq \mathbb{P}^2_{\mathbb{C}}$ parameterize the intersection $C \cap D$. The point $(\alpha : \beta : \gamma)^T \in C \cap D$ is a *k-fold intersection point* if the corresponding zero $(\alpha : \beta)^T$ of r has order k.

Theorem 8.27 (Bézout's Theorem) *If two projective curves $C, D \subseteq \mathbb{P}^2_{\mathbb{C}}$ of degrees n and m have no common component, then the sum of the multiplicities of their intersection points is nm.*

Proof Using the above notation let s be the cardinality of the intersection $C \cap D$. For $p_i \in C \cap D$ let k_i denote the intersection multiplicity. Then we have $\sum_{i=1}^{s} k_i = \deg r = nm$. $\square$

8.7 Algebraic Curves Using **`Maple`**

Many computer algebra systems enable us to study and visualize algebraic curves. We think it is sufficient to illustrate this with the commercial universal software package `Maple`.

The following `Maple` commands load packages for studying and illustrating algebraic curves.

```
> with(algcurves):
> with(plots):
```

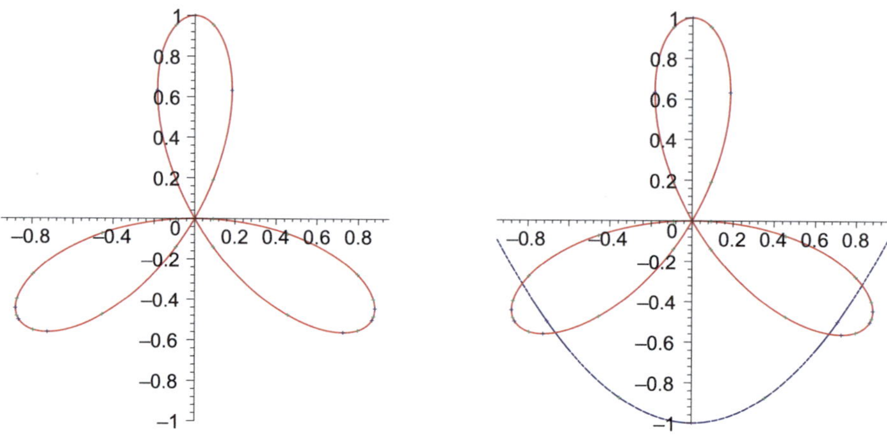

Fig. 8.8 A three-leaf clover and its intersection with a parabola

We define the polynomial $f := (x^2 + y^2)^2 + 3x^2y - y^3 \in \mathbb{C}[x, y]$ and visualize the affine curve C defined by f via

```
> f := (x^2+y^2)^2 + 3*x^2*y - y^3;
> plot_real_curve(f,x,y);
```

See the left hand side of Fig. 8.8. The curve C of degree 4 is also called a *three-leaf clover*. We will study now the parabola given by

```
> g := y-(x^2-1);
```

see the right hand side of Fig. 8.8.

To calculate the x-coordinates of the real intersection points of C with the parabola defined by g we use

```
> r := resultant(f,g,y);
```

to determine the resultant $\mathrm{Res}_y(f, g)$, which gives

```
    r := 9*x^4-8*x^2+2-3*x^6+x^8.
```

Using the command

```
> fsolve(r,x,complex);
```

we get a numerical approximation of the complex zeros of r. As seen in Fig. 8.9, r has real zeros with the numerical values

$$\pm 0.8281 \quad \text{and} \quad \pm 0.6656.$$

Additionally, we have four non-real zeros

$$\pm 1.3232 \pm 0.9029i.$$

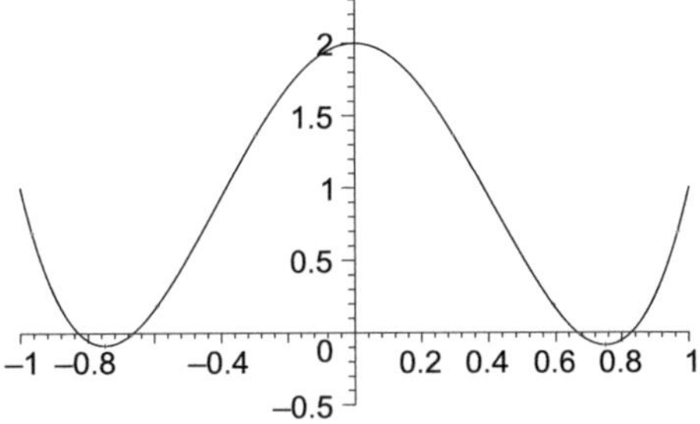

Fig. 8.9 The graph of the resultant r

Theorem 8.8 implies that for every zero α of r, the univariate polynomials $f(\alpha, y)$ and $g(\alpha, y)$ have a common factor and therefore a common zero. Hence we get for every α (at least) one point in the intersection of C with the parabola.

When plotting the resultant r with `Maple` a "naive" call of the function `plot` does not yield a reasonable result, since $\deg r = 8$ is relatively large. We can fix this by either choosing the number of points at which the function is evaluated to be large (optional argument `numpoints`) or by using the more intelligent function `plot_real_curve`.

8.8 Exercises

Exercise 8.28 Let

$$f = a_n x^n + a_{n-1} x^{n-1} + \cdots + a_0,$$

$$g = b_m x^m + b_{m-1} x^{m-1} + \cdots + b_0$$

be polynomials in $\mathbb{C}[x]$ of degrees n and m with their (not necessarily pairwise distinct) zeros $\alpha_1, \ldots, \alpha_n$ and $\beta_1, \ldots, \beta_m$.

(a) Show

$$\mathrm{Res}(f, g) = a_n^n b_m^m \prod_{i=1}^{n} \prod_{j=1}^{m} (\alpha_i - \beta_j)$$

$$= a_n^n \prod_{i=1}^{n} g(\alpha_i) = (-1)^{mn} b_m^m \prod_{j=1}^{m} f(\beta_j).$$

(b) Deduce that $\mathrm{Res}(f_1 f_2, g) = \mathrm{Res}(f_1, g)\,\mathrm{Res}(f_2, g)$ for $f_1, f_2, g \in \mathbb{C}[x]$. Is this statement true only in $\mathbb{C}$ or over other fields as well?

Exercise 8.29 Show that the set $A = \{(x, x) \in \mathbb{R}^2 : x \geq 0\}$ is not an algebraic hypersurface and cannot be written as an intersection of hypersurfaces.

Exercise 8.30 For which $\alpha, \beta \in \mathbb{R}$ are all points in $V(f) \cap V(g)$ with

$$f(x, y) = (x - 2\alpha)^2 + y^2 - 1, g(x, y) = (x - 2\beta)^2 + y^2 - 1 \in \mathbb{C}[x, y]$$

real?

Exercise 8.31 Let $f \in \mathbb{C}[x_1, \ldots, x_n]$ be a non-constant polynomial with homogenization $\bar{f} \in \mathbb{C}[x_0, x_1, \ldots, x_n]$. Show that f is irreducible if and only if $\bar{f}$ is irreducible.

8.9 Remarks

For the numerical computation of eigenvalues we refer to the textbook of Stoer and Bulirsch [93]. The fundamentals of Galois theory can be found in Howie's book [66].

It is extremely difficult to accurately illustrate algebraic surfaces including their singularities. One interesting possibility is to use ray-tracing techniques, as is shown, for example, by `surfex` [64]. The examples of Fig. 8.4 were generated using `SingSurf` [79] and `JavaView` [82]. `SingSurf` provides a grid model of the surface and `JavaView` provides an interactive view of the model.

Bézout's Theorem can be traced back to the 18th century (however, the proof given by Étienne Bézout was incorrect, and it appeared after the first proof of the theorem). Our approach is based on Fischer's book [39]. Furthermore, we would like to refer to the books of Cox, Little, O'Shea [29] and Kirwan [70].

Chapter 9
Gröbner Bases and Buchberger's Algorithm

We next examine the problem of finding the common roots of a finite set of polynomials over a field K. To do this, we first introduce some necessary algebraic structures. *Gröbner bases* play a key role in the computational aspect of this problem.

In Chapter 10 we will see how to computationally solve arbitrary systems of polynomial equations using Gröbner bases.

9.1 Ideals and the Univariate Case

In the following we study a polynomial ring over an arbitrary field K. In Chapter 8 we defined (affine and projective) algebraic varieties for a given polynomial. We now generalize this definition to a set of polynomials. Let $S \subseteq K[x_1, \ldots, x_n]$ be an arbitrary set of polynomials. Then

$$\mathrm{V}(S) := \left\{ a \in K^n : f(a_1, \ldots, a_n) = 0 \text{ for all } f \in S \right\} = \bigcap_{f \in S} \mathrm{V}(f)$$

is called the *affine variety* of S over the field K. Hence, an affine variety is an intersection of affine hyperplanes. One can immediately observe that any common root of the polynomials $f_1, \ldots, f_t \in K[x_1, \ldots, x_n]$ is also a root of $\sum_{i=1}^t h_i f_i$. This holds for an arbitrary choice of $h_1, \ldots, h_t \in K[x_1, \ldots, x_n]$, which motivates the following definition.

Definition 9.1 A non-empty set $I \subseteq K[x_1, \ldots, x_n]$ is called an *ideal* if for all $f, g \in I$ and all $h \in K[x_1, \ldots, x_n]$ we have $f + g \in I$ and $hf \in I$.

M. Joswig, T. Theobald, *Polyhedral and Algebraic Methods in Computational Geometry*,   137
Universitext, DOI 10.1007/978-1-4471-4817-3_9,

For $S \subseteq K[x_1, \ldots, x_n]$ we denote by $\langle S \rangle$ the *ideal generated by* S, i.e., the smallest ideal of $K[x_1, \ldots, x_n]$ that contains S. We have

$$\langle S \rangle = \left\{ \sum_{i=1}^{t} h_i f_i : f_1, \ldots, f_t \in S, \, h_1, \ldots, h_t \in K[x_1, \ldots, x_n], \, t \in \mathbb{N} \right\}.$$

The following exercise illustrates how varieties can be defined using ideals.

Exercise 9.2 Show that $\mathrm{V}(S) = \mathrm{V}(\langle S \rangle)$.

A generating system of an ideal I is also called a *basis* of I. Here we need to stress that—unlike in the case of vector spaces—an ideal can have bases of different cardinalities: For example, every subset of an ideal I which contains a basis of I is also a basis of I. In Corollary 9.23, which is also known as the *Hilbert Basis Theorem*, we will see that every ideal $I \subseteq K[x_1, \ldots, x_n]$ is finitely generated.

Not every basis of an ideal is of equal quality. Some bases allow for the observation of more characteristics of the ideal than others. We illustrate this with an example.

Example 9.3 Let $f = x^2 y + x + 1$, $g = x^3 y + x + 1 \in \mathbb{C}[x, y]$. In order to compute the common roots of f and g, it is helpful to have a polynomial of $I = \langle f, g \rangle$ that depends on only one unknown (e.g. on x). In this case we have

$$x^2 - 1 = x \cdot f - g \in I.$$

Therefore, for every common root $(a, b)^T$ of f, g we know that $a \in \{-1, 1\}$. Substituting and solving the equations for y shows that the two points $(-1, 0)^T$ and $(1, -2)^T$ are the common roots of f and g. Figure 9.1 illustrates the real part of the curves $\mathrm{V}(f)$ and $\mathrm{V}(g)$. We have $x \cdot f - (x^2 - 1) = g$, hence $I = \langle f, x^2 - 1 \rangle$.

At this point, we briefly remark that the resultant $\mathrm{Res}_y(f, g) = -x^2(x^2 - 1)$ is also contained in I. We shall return to this connection in Chapter 10 (Proposition 10.4).

The previous example suggests the idea of solving a system of polynomial equations via step by step elimination of variables followed by backwards substitution. This corresponds to solving a linear system of equations in row echelon form. This approach motivates the following term: For an ideal $I = \langle f_1, \ldots, f_t \rangle \subseteq K[x_1, \ldots, x_n]$ and $i \in \{1, \ldots, n - 1\}$ let

$$I \cap K[x_{i+1}, \ldots, x_n]$$

denote the i-th *elimination ideal*.

Exercise 9.4 Show that the i-th elimination ideal of I is indeed an ideal in $K[x_{i+1}, \ldots, x_n]$.

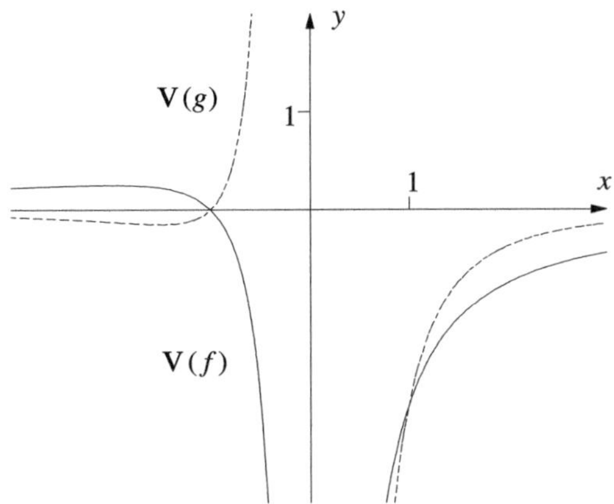

Fig. 9.1 Varieties V(f) and V(g)

We now lay the foundation for the study of elimination ideals in Chapter 10. To do this we study the question of how, given an ideal I and a polynomial f, we can determine if f is in I. This is the so-called *ideal membership problem* for which Algorithm 9.3 on p. 148 provides a solution.

We first examine the special case of the ideal membership problem with one unknown: For given polynomials $f_1, \ldots, f_t, f \in K[x]$ we ask whether $f \in \langle f_1, \ldots, f_t \rangle$. A polynomial ring $K[x]$ in one variable is a *Euclidean ring* since we can define a division algorithm. Division (via the Euclidean Algorithm 9.1) allows us to compute the greatest common divisor g of the polynomials $f_1, \ldots, f_t$, with $\langle g \rangle = \langle f_1, \ldots, f_t \rangle$. Furthermore, the Euclidean algorithm allows us to solve the ideal membership problem, since it in particular enables us to determine if the remainder of f divided by g is 0.

We assume that the reader knows the basic principles of the algorithm. However, due to the significance of these two algorithms in our further work, we will illustrate them.

For two polynomials $f, g \in K[x] \setminus \{0\}$ there exist $r, s \in K[x]$ such that

$$f = q \cdot g + r \quad \text{where } \deg r < \deg g. \tag{9.1}$$

When $\deg f \geq \deg g$, we do the following: Assume that $f = \sum_{i=0}^{n} a_i x^i$ and $g = \sum_{j=0}^{m} b_j x^j$ where $n \geq m$ and $a_n, b_m \neq 0$. Via induction over the degree we can assume that the polynomial $h := f - \frac{a_n}{b_m} \cdot x^{n-m} \cdot g$ of degree $\leq n - 1$ has a decomposition $h = q' \cdot g + r$, such that $\deg r < \deg g$. This implies

$$f = h + \frac{a_n}{b_m} \cdot x^{n-m} \cdot g = \left(q' + \frac{a_n}{b_m} x^{n-m} \right) g + r.$$

Using $q := q' + \frac{a_n}{b_m} x^{n-m}$ we get the desired statement. We denote the *remainder r* as $\mathrm{rem}(f; g)$ and write $g \mid f$ if $\mathrm{rem}(f; g) = 0$.

Definition 9.5 Let K be a field. A polynomial $g \in K[x]$ is called a *greatest common divisor* (gcd) of $f_1, \ldots, f_t \in K[x] \setminus \{0\}$ if the following conditions are satisfied.

Algorithm 9.1: The Euclidean algorithm

Input: $f, g \in K[x] \setminus \{0\}$ with $\deg f \geq \deg g$
Output: $\gcd(f, g)$

1   $r_0 \leftarrow f; r_1 \leftarrow g; i \leftarrow 1$
2   **while** $r_i \neq 0$ **do**
3      $r_{i+1} \leftarrow \mathrm{rem}(r_{i-1}; r_i)$
4      $i \leftarrow i + 1$

5   **return** r_{i-1}

(a) $g \mid f_i$ for all $i \in \{1, \ldots, t\}$;
(b)   if $h \mid f_1, \ldots, h \mid f_t$ then $h \mid g$ for all $h \in K[x]$.

In every unique factorization domain there exists a greatest common divisor which is unique up to multiplication by a unit (here a non-zero constant in K), see Appendix A. For uniqueness we choose the gcd with leading coefficient 1.

Analogously we can define the *least common multiple* of $f_1, \ldots, f_t$. Alternatively we can read the following computational rule for two polynomials as a definition:

$$\mathrm{lcm}(f_1, f_2) := \frac{f_1 f_2}{\gcd(f_1, f_2)}.$$

This also shows that the computation of the least common multiple can be reduced to the computation of the greatest common divisor.

A special property of the ring $K[x]$, or of any Euclidean ring, is that the gcd can be algorithmically computed.

The Euclidean Algorithm 9.1 terminates since the degrees of the polynomials r_i are strictly decreasing. We denote by q_i the polynomial such that in Step 2 we have

$$r_{i-1} = q_i \cdot r_i + r_{i+1}. \tag{9.2}$$

To prove that the algorithm is correct we show that $r := r_{i-1}$, which is returned in the last step, satisfies the two conditions from Definition 9.5. Using (9.2) we can successively deduce that r divides the remainders $r_{i-2}, r_{i-3}, \ldots, r_1 = g$ and $r_0 = f$. If h divides f as well as g, then h divides $r_2, r_3, \ldots, r_{i-1}$. This can also be deduced from (9.2).

Every remainder computed throughout the Euclidean algorithm is contained in the ideal $\langle f, g \rangle$ of the two input polynomials $f, g \in K[x]$ and hence we have $\gcd(f, g) \in \langle f, g \rangle$. Therefore,

$$\langle \gcd(f, g) \rangle = \langle f, g \rangle.$$

Example 9.6 Applying the Euclidean algorithm to the two polynomials $f = x^4 - x^3$ and $g = x^3 - x$ repeatedly yields $(q_1, r_2) = (x - 1, x^2 - x)$ and $(q_2, r_3) = (x + 1, 0)$, so that $x^2 - x$ is the gcd of f and g.

To determine a single generator of an ideal which is given by more than two polynomials it is sufficient to verify the following rule:

Exercise 9.7 For $t \geq 3$ we have $\gcd(f_1, \ldots, f_t) = \gcd(f_1, \gcd(f_2, \ldots, f_t))$.

Given a sequence of generators $f_1, \ldots, f_t$ of an ideal $I \subseteq K[x]$, the representation of I as a principal ideal $I = \langle \gcd(f_1, \ldots, f_t) \rangle$ is called a *normal form* of I. This notion is justified by the following exercise.

Exercise 9.8 For univariate polynomials $f_1, \ldots, f_t, g_1, \ldots, g_s \in K[x] \setminus \{0\}$ with $\langle f_1, \ldots, f_t \rangle$ equal to $\langle g_1, \ldots, g_s \rangle$, show that up to a constant factor the polynomials $\gcd(f_1, \ldots, f_t)$ and $\gcd(g_1, \ldots, g_s)$ coincide.

In particular, the special case $s = 1$ implies that in any representation of an ideal $I \subseteq K[x]$ as a principal ideal $I = \langle g_1 \rangle$, the polynomial g_1 is uniquely determined up to a constant factor.

The Euclidean algorithm serves to compute a normal form for a given ideal I in $K[x]$. If an ideal in $K[x]$ is given in normal form, i.e., by a single generator, then the Euclidean division solves the ideal membership problem. The goal of the following sections is to generalize these two methods to polynomial rings with an arbitrary number of unknowns.

9.2 Monomial Orders

The degree naturally defines a partial order on the polynomials in one unknown, which were studied in the previous section. The remainder polynomial, which is the result of the division of a polynomial f by g, is smaller than g with respect to this partial order. To define a proper division in the multivariate case it is necessary to first define a suitable order on the set of monomials.

A monomial $x_1^{\alpha_1} \cdots x_n^{\alpha_n}$ in $K[x_1, \ldots, x_n]$ is denoted by x^α, where $\alpha = (\alpha_1, \ldots, \alpha_n) \in \mathbb{N}^n$ is a multi-index. In Definition 8.14 we defined the total degree of a monomial as $\operatorname{tdeg} x^\alpha := \alpha_1 + \cdots + \alpha_n$. The notation $|\alpha|$ is also used as an alternative to $\operatorname{tdeg} x^\alpha$.

Definition 9.9 A *monomial order* on $K[x_1, \ldots, x_n]$ is a relation $\prec$ on $\mathbb{N}^n$ (or equivalently a relation on the set of monomials x^α for $\alpha \in \mathbb{N}^n$), which satisfies the following properties.

(a) The relation $\prec$ is a well-ordered relation on $\mathbb{N}^n$, i.e., every non-empty subset of $\mathbb{N}^n$ has a minimal element with respect to $\prec$.
(b) $\alpha \prec \beta$ and $\gamma \in \mathbb{N}^n$ implies $\alpha + \gamma \prec \beta + \gamma$.

Every well-ordered relation is a total order. From condition (b) it follows that the zero vector (respectively the empty monomial 1) is the unique smallest element with respect to every monomial order. The second condition requires a *compatibility with respect to multiplication*: $x^\alpha \cdot x^\gamma \prec x^\beta \cdot x^\gamma$ (when expressed in the monomial description).

Definition 9.10 (Lexicographic order)   Let $\alpha, \beta \in \mathbb{N}^n$. We define $x^\alpha \prec_{\text{lex}} x^\beta$ if the leftmost non-zero coefficient in the difference $\beta - \alpha \in \mathbb{Z}^n$ is positive.

Example 9.11 We have $(4, 3, 1) \succ_{\text{lex}} (3, 7, 10)$ and $(4, 3, 1) \prec_{\text{lex}} (4, 7, 10)$. Expressed as monomials in $K[x, y, z]$, this translates to $x^4 y^3 z^1 \succ_{\text{lex}} x^3 y^7 z^{10}$ and $x^4 y^3 z^1 \prec_{\text{lex}} x^4 y^7 z^{10}$ respectively.

The relation $\prec_{\text{lex}}$ is a monomial order. It suffices to check that the relation is a well-ordered relation. If we assume that $\prec_{\text{lex}}$ is not a well-ordered relation, then we can find a strictly decreasing series

$$\alpha^{(1)} \succ_{\text{lex}} \alpha^{(2)} \succ_{\text{lex}} \alpha^{(3)} \succ_{\text{lex}} \cdots \tag{9.3}$$

of elements in $\mathbb{N}^n$. By the definition of the lexicographic order, the leftmost entries $\alpha_1^{(i)}$ define a non-increasing series in $\mathbb{N}$. Since the set of natural numbers is well-ordered, there exists an N_1 such that $(\alpha^{(i)})_1 = (\alpha^{(N_1)})_1$ for all $i \geq N_1$. Now by considering only the series elements after the index N_1, we can in the same way deduce that there exist $N_2, \ldots, N_n$ such that $(\alpha^{(i)})_j = (\alpha^{(N_j)})_j$ for all $i \geq N_j$ and $j \in \{2, \ldots, n\}$. This contradicts the series (9.3) being strictly decreasing.

A monomial order yields a unique sorted description for arbitrary polynomials. For the remaining part of this section we will fix a monomial order $\prec$ on $K[x_1, \ldots, x_n]$. For a non-zero polynomial $f = \sum_\alpha c_\alpha x^\alpha$ in $K[x_1, \ldots, x_n]$ let $\alpha^* := \max_\prec \{\alpha : c_\alpha \neq 0\}$. The *leading monomial* of f is $\text{lm}_\prec(f) := x^{\alpha^*}$ and the corresponding coefficient $\text{lc}_\prec(f) := c_{\alpha^*}$ is called the *leading coefficient*. Their product

$$\text{lt}_\prec(f) := \text{lc}_\prec(f) \cdot \text{lm}_\prec(f) = c_{\alpha^*} \cdot x^{\alpha^*}$$

is called the *leading term* of f. When the monomial order is contextually clear it is often neglected in the notation.

Example 9.12 For $f = 5x^4 y^3 z + 2x^3 y^7 z^{10}$ in $K[x, y, z]$ we have that $\text{lt}_{\prec_{\text{lex}}}(f) = x^4 y^3 z$, $\text{lc}_{\prec_{\text{lex}}}(f) = 5$ and $\text{lm}_{\prec_{\text{lex}}}(f) = 5x^4 y^3 z$ with respect to the lexicographic order.

We are now able to generalize the division algorithm to the multivariate case. Here there is a major difference in comparison to the univariate case: It is useful to describe the division of a polynomial $f \in K[x_1, \ldots x_n]$ by a set of polynomials $(f_1, \ldots, f_t)$ since ideals in $K[x_1, \ldots, x_n]$ are in general not generated by a single polynomial.

We look at the leading monomial $\text{lm}(f)$ of f and check if division by any of the leading monomials $\text{lm}(f_1), \ldots, \text{lm}(f_t)$ results in a remainder of 0. For the first

---

Algorithm 9.2: The multivariate division algorithm

Input: $f, f_1, \ldots, f_t \in K[x_1, \ldots, x_n]$ with $f_i \neq 0$
Output: $a_1, \ldots, a_t, r$ with $f = \sum_{i=1}^{t} a_i f_i + r$

1   $a_i \leftarrow 0$ for all $i \in \{1, \ldots, t\}$
2   $p \leftarrow f$
3   **while** $p \neq 0$ **do**
4      $m \leftarrow \mathrm{lt}(p)$
5      $i \leftarrow 1$
6   **while** $i \leq t$ and $m \neq 0$ **do**
7   **if** $\mathrm{lt}(f_i)$ divides m **then**
8            $a_i \leftarrow a_i + \frac{m}{\mathrm{lt}(f_i)}$; $p \leftarrow p - \frac{m}{\mathrm{lt}(f_i)} f_i$
9            $m \leftarrow 0$
10         $i \leftarrow i + 1$
11      $r \leftarrow r + m$; $p \leftarrow p - m$
12 **return** $(a_1, \ldots, a_t; r)$

---

polynomial f_k which satisfies this condition, we subtract a suitable multiple of f_k from f,

$$f - \frac{\mathrm{lt}(f)}{\mathrm{lt}(f_k)} f_k,$$

and obtain a new polynomial which is strictly smaller than f with respect to the monomial order. We replace f by the new polynomial and repeat the process. If the leading monomial of f is not divisible by any of the leading terms $\mathrm{lt}(f_1), \ldots, \mathrm{lt}(f_t)$, we add the leading term to the remainder, subtract it from f and start again at the beginning.

The remainder r which is produced by Algorithm 9.2 is called the *remainder* of f after division by $(f_1, \ldots, f_t)$ and we denote it by $\mathrm{rem}(f; f_1, \ldots, f_t)$. In general this remainder is not independent of the order of the polynomials by which we divide.

Example 9.13 Let $f = xy^2 - y$, $f_1 = xy - 1$ and $f_2 = y^2 + 1$ be polynomials in $K[x, y]$. With respect to the lexicographic order and the ordering (f_1, f_2) of the polynomials, the division algorithm divides the leading term xy^2 by xy resulting in y. Since $f - y \cdot f_1 = 0$ the algorithm terminates and returns the decomposition

$$xy^2 - y = y \cdot (xy - 1) + 0 \cdot (y^2 + 1) + 0.$$

If we reverse the ordering of the polynomials, i.e., we divide by (f_2, f_1), the term xy^2 is divided by the leading monomial y^2, resulting in x. Since the polynomial $f - x \cdot f_2 = -y - x$ is not divisible any further by f_1 or f_2, the algorithm yields the decomposition

$$f = x \cdot \left(y^2 + 1\right) + 0 \cdot (xy - 1) + (-x - y).$$

Our remainders are: $\mathrm{rem}(f; f_1, f_2) = 0$ and $\mathrm{rem}(f; f_2, f_1) = -x - y$.

In general, the multivariate division algorithm results in a representation of the following form.

Lemma 9.14 *For given polynomials* $f, f_1, \ldots, f_t \in K[x_1, \ldots, x_n]$ *the Division Algorithm 9.2 returns polynomials* $a_1, \ldots, a_t$ *and* $r = \mathrm{rem}(f; f_1, \ldots, f_t)$, *for which we have*

$$f = a_1 f_1 + \cdots + a_t f_t + r,$$

where no term of r *is divisible by any of the monomials* $\mathrm{lm}(f_1), \ldots, \mathrm{lm}(f_t)$. *Furthermore, we have for each* $i \in \{1, \ldots, t\}$ *with* $a_i \neq 0$ *that*

$$\mathrm{lm}(a_i f_i) \preceq \mathrm{lm}(f).$$

Proof It is clear by the construction of the algorithm that no term of the remainder r is divisible by any of the leading monomials $\mathrm{lm}(f_1), \ldots, \mathrm{lm}(f_t)$. The assignment $a_i \leftarrow a_i + \frac{\mathrm{lt}(p)}{\mathrm{lt}(f_i)}$ ensures that the product $a_i \, \mathrm{lt}(f_i)$ is a sum of terms of f. However, the terms of f are dominated by their leading term. $\qquad\square$

Exercise 9.15 Let $\prec$ be a monomial order on $K[x_1, \ldots, x_n]$. Show that

$$\alpha \prec^{\mathrm{tdeg}} \beta \quad :\Longleftrightarrow \quad \mathrm{tdeg}\,\alpha < \mathrm{tdeg}\,\beta \quad \text{or} \quad (\mathrm{tdeg}\,\alpha = \mathrm{tdeg}\,\beta \text{ and } \alpha \prec \beta),$$

defines a monomial order.

The construction in Exercise 9.15 can, in certain cases, yield a monomial order even if the original order does not satisfy all the axioms of a monomial order. The next exercise exhibits this phenomenon for $\prec_{\mathrm{grevlex}}$, a monomial order that is often a very efficient one in practical computations.

Exercise 9.16 Let $\alpha, \beta \in \mathbb{N}^n$. We define $x^\alpha <_{\mathrm{revlex}} x^\beta$ if the rightmost non-zero coefficient in the difference $\beta - \alpha \in \mathbb{Z}^n$ is negative.

(a) Show that the *reverse lexicographic order* $<_{\mathrm{revlex}}$ is not a monomial order.
(b) Show that the *graded reverse lexicographic order* defined by

$$\alpha \prec_{\mathrm{grevlex}} \beta \quad :\Longleftrightarrow \quad \mathrm{tdeg}\,\alpha < \mathrm{tdeg}\,\beta \quad \text{or}$$
$$(\mathrm{tdeg}\,\alpha = \mathrm{tdeg}\,\beta \text{ and } \alpha <_{\mathrm{revlex}} \beta),$$

is a monomial order.

9.3 Gröbner Bases and the Hilbert Basis Theorem

In this section we introduce the key concept for solving the ideal membership problem. We start with an example that illustrates why the multivariate case is much more complicated than the univariate case.

Example 9.17 Let $f_1 = xy + 1$, $f_2 = yz + 1$ be polynomials in $K[x, y]$. In the univariate case it would be desirable to use Euclidean division to determine if the polynomial $f = z - x$ is contained in the ideal $I = \langle f_1, f_2 \rangle$. In fact we do have

$$z - x = z \cdot (xy + 1) - x(yz + 1) \in \langle f_1, f_2 \rangle.$$

However, neither for the ordering (f_1, f_2) nor for the ordering (f_2, f_1) is the remainder zero when applying Euclidean division with respect to lexicographic order. Of course, adding $z - x$ to the ideal basis would give that division of $z - x$ by the new basis would have 0 as the remainder.

It may seem naive to enlarge the original generating system of an ideal by proper polynomials so that *every* polynomial of the ideal has a remainder of zero when divided by the basis. However, this can be algorithmically achieved. What we need for this is a criterion that determines if the generating system is large enough.

We denote the set of leading terms of an ideal I with respect to the monomial order $\prec$ by $\mathrm{lt}_\prec(I)$. The ideal $\langle \mathrm{lt}_\prec(I) \rangle$ generated by the leading terms is called the *initial ideal* of I with respect to $\prec$, and we write $\mathrm{in}_\prec(I) := \langle \mathrm{lt}_\prec(I) \rangle$.

Definition 9.18 Let I be an ideal. A finite subset $G = \{g_1, \ldots, g_t\} \subseteq I$ is called a *Gröbner basis* of I with respect to the monomial order $\prec$ if the leading terms $\mathrm{lt}_\prec(g_1), \ldots, \mathrm{lt}_\prec(g_t)$ generate the initial ideal of I, i.e.,

$$\left\langle \mathrm{lt}_\prec(g_1), \ldots, \mathrm{lt}_\prec(g_t) \right\rangle = \mathrm{in}_\prec(I).$$

Our next important intermediate goal is to show that every ideal has a Gröbner basis. We begin by proving this statement for the special case of monomial ideals. *Monomial ideals* are those ideals which have a generating system consisting only of monomials. Initial ideals are always monomial ideals.

Lemma 9.19 *Let $I = \langle x^\alpha : \alpha \in A \rangle$ where $A \subseteq \mathbb{N}^n$ is a monomial ideal. We have $x^\beta \in I$ if and only if x^β is a multiple of x^α for an $\alpha \in A$.*

Proof If x^β is a multiple of x^α for an $\alpha \in A$, then by the definition of an ideal, $x^\beta \in I$.

Conversely, if $x^\beta \in I$, then there exists a representation $x^\beta = \sum_{i=1}^{t} h_i x^{\alpha^{(i)}}$ with $h_i \in K[x_1, \ldots, x_n]$ and $\alpha^{(i)} \in A$ for $1 \leq i \leq t$. Every term of the polynomial on the right hand side of the equation is a multiple of a term x^α for some $\alpha \in A$. Therefore, the polynomial on the left hand side of the equation also has this property. $\square$

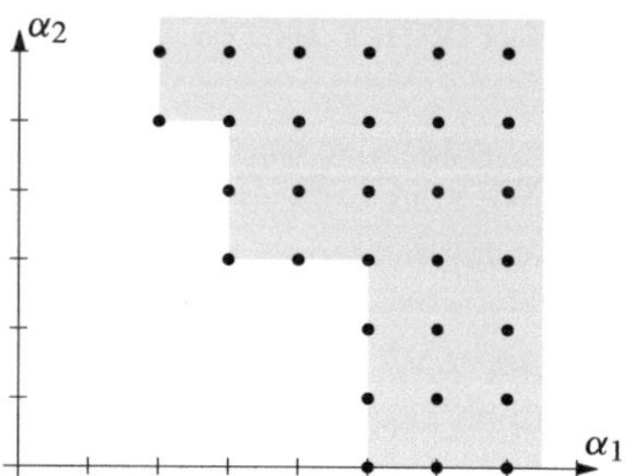

Fig. 9.2 A visualization of the Gordan–Dickson lemma for $n = 2$. Every lattice point (α_1, α_2) represents a monomial $x_1^{\alpha_1} x_2^{\alpha_2}$

The following theorem shows that monomial ideals are finitely generated.

Theorem 9.20 (Gordan–Dickson Lemma) *Every non-empty set M of monomials in $K[x_1, \ldots, x_n]$ contains a finite subset $E \subseteq M$ such that every monomial of M is a multiple of a monomial in E.*

Before beginning the proof, we illustrate the theorem for the case $n = 2$. Each point (i, j) in Fig. 9.2 represents a monomial $x^i y^j$ in $K[x, y]$. If a monomial $x^i y^j$ is contained in a monomial ideal I, then Lemma 9.19 states that every monomial $x^k y^l$ with $k \geq i$ and $l \geq j$ is contained in I as well. So the Gordan–Dickson lemma implies that the points corresponding to monomials in I can be represented as a finite union of transposed copies of the points in the positive orthant.

Proof The proof is by induction over the number of unknowns n. For $n = 1$ we have $M = \{x^\alpha : \alpha \in A\}$ for a subset $A \subseteq \mathbb{N}$. A has a smallest element β. Using Lemma 9.19 we conclude $I = \langle x^\beta \rangle$.

So let $n \geq 2$ and assume that the statement is true for $n - 1$ unknowns. Take an arbitrary monomial

$$x^\alpha = x_1^{\alpha_1} \cdots x_n^{\alpha_n}$$

from M.

We first show that every monomial $x^\beta \in M$ which is not a multiple of x^α belongs to at least one of the sets $M_{i,j}$, where: for $i \in \{1, \ldots, n\}$ and $j \in \{0, \ldots, \alpha_i - 1\}$, $M_{i,j}$ is the set of those monomials $x^\gamma \in M$ for which $\deg_{x_i}(x^\gamma) = j$. Since x^α does not divide the monomial x^β, we have $\beta_i < \alpha_i$ for some $i \in \{1, \ldots, n\}$. Hence, $x^\beta \in M_{i,\beta_i}$.

Let $M'_{i,j}$ be the set of monomials in $K[x_1, \ldots, x_{i-1}, x_{i+1}, \ldots, x_n]$ that can be obtained from monomials of $M_{i,j}$ by dropping the factor x_i^j. By the inductive hypothesis there exist finite subsets $E'_{i,j} \subseteq M'_{i,j}$ such that every monomial in $M'_{i,j}$ is a multiple of the monomial $E'_{i,j}$. We define

$$E_{i,j} := \left\{ p \cdot x_i^j : p \in E'_{i,j} \right\}.$$

Now it is clear that every monomial in M is a multiple of a monomial in the finite set

$$E := \{x^\alpha\} \cup \bigcup_{i,j} E_{i,j} \subseteq M.$$

$\square$

Remark 9.21 This lemma will play a key role in proving the termination of several algorithms. The statement is actually purely combinatorial: Given a set $\mathcal{A}$ of subsets of $\mathbb{N}^n$ such that every $A \in \mathcal{A}$ is of the form $\alpha_A + \mathbb{N}^n$ with $\alpha_A \in \mathbb{N}^n$, the union $\bigcup_{A \in \mathcal{A}} A$ is a finite union, i.e., there exist $A_1, \ldots, A_k \in \mathcal{A}$ with $\bigcup_{A \in \mathcal{A}} A = \bigcup_{i=1}^{k} A_i$.

Using the Gordan–Dickson lemma it is now possible to prove that every non-zero ideal in $K[x_1, \ldots, x_n]$ has a Gröbner basis.

Theorem 9.22 *Let $\prec$ be a monomial order on $K[x_1, \ldots, x_n]$. Then:*

(a) *Every non-zero ideal I has a Gröbner basis.*
(b) *The elements of a Gröbner basis of I generate the ideal I.*

Proof Let $I \neq \{0\}$ be an ideal.

(a): The initial ideal $\mathrm{in}_\prec(I)$ is generated by the monomials $\mathrm{lm}_\prec(g)$, with $g \in I \setminus \{0\}$. By the Gordan–Dickson lemma 9.20 there exist finitely many $g_1, \ldots, g_t$ with

$$\langle \mathrm{lt}_\prec(g_1), \ldots, \mathrm{lt}_\prec(g_t) \rangle = \mathrm{lt}_\prec(I),$$

which ensures the existence of a Gröbner basis.

(b): The ideal J which is generated by the polynomials $g_1, \ldots, g_t$ of a Gröbner basis is clearly contained in I. To show the reverse inclusion we assume that $I \setminus J \neq \emptyset$. Let f be a polynomial in $I \setminus J$ with a leading term that is minimal with respect to $\prec$. Since $\mathrm{lm}_\prec(g_1), \ldots, \mathrm{lm}_\prec(g_t)$ generate the initial ideal $\mathrm{in}_\prec(I)$, there exist polynomials $h_1, \ldots, h_t$ with

$$\mathrm{lm}_\prec(f) = \mathrm{lm}_\prec(g_1) \cdot h_1 + \cdots + \mathrm{lm}_\prec(g_t) \cdot h_t.$$

The polynomial

$$g = f - \sum_{i=1}^{t} g_i h_i$$

is contained in I but not in J (otherwise we would have $f \in J$). We also have that the leading monomial of f does not appear in g, which means that the corresponding coefficient is zero. Hence $\mathrm{lm}_\prec(g)$ is smaller than $\mathrm{lm}_\prec(f)$ with respect to the monomial order $\prec$. This contradicts the minimality of f. We therefore have $I = J$, which proves our statement.

$\square$

As an immediate consequence of Theorem 9.22 we get the following finiteness statement.

Algorithm 9.3: A solution of the ideal membership problem

Input: $f, g_1, \ldots, g_t \in K[x_1, \ldots, x_n]$, such that $G := \{g_1, \ldots, g_t\}$ is a Gröbner
basis of the ideal $I = \langle G \rangle$ with respect to the monomial order $\prec$

Output: Determine if $f \in I$

1 $r \leftarrow \mathrm{rem}_\prec(f; g_1, \ldots, g_t)$

2 **if** $r = 0$ **then**

3 $\quad$ **return** "Yes"

4 **else**

5 $\quad$ **return** "No"

Corollary 9.23 (Hilbert Basis Theorem) *Every ideal $I \subseteq K[x_1, \ldots, x_n]$ has a finite generating system.*

The important property of Gröbner bases is that they provide a solution to the ideal membership problem, as carried out in Algorithm 9.3.

Correctness of Algorithm 9.3 If $\mathrm{rem}_\prec(f; g_1, \ldots, g_t) = 0$, then f is contained in I. It remains to be shown that $\mathrm{rem}_\prec(f; g_1, \ldots, g_t) \neq 0$ implies that $f \notin I$. Assume that $\mathrm{rem}_\prec(f; g_1, \ldots, g_t) \neq 0$ and $f \in I$. Then the remainder $r = \mathrm{rem}_\prec(f; g_1, \ldots, g_t) \in I$ and therefore $\mathrm{lt}_\prec(r) \in \mathrm{in}_\prec(I)$. Since G is a Gröbner basis, it follows that $\mathrm{in}_\prec(I) = \langle \mathrm{lt}_\prec(g_1), \ldots, \mathrm{lt}_\prec(g_t) \rangle$. By Lemma 9.19, $\mathrm{lt}_\prec(r)$ is a multiple of a leading term $\mathrm{lt}_\prec(g_i)$ for an $i \in \{1, \ldots, t\}$. But by Lemma 9.14, the divisibility of $\mathrm{lt}_\prec(r)$ by $\mathrm{lt}_\prec(g_i)$ contradicts the fact that r is a remainder of the division by $g_1, \ldots, g_t$. $\square$

For the remaining part of this section assume that $G = \{g_1, \ldots, g_t\}$ is a Gröbner basis of the ideal $I \subseteq K[x_1, \ldots, x_n]$ with respect to the monomial order $\prec$.

Exercise 9.24 Show that Euclidean division is independent of the order of polynomials in G:

$$\mathrm{rem}_\prec(f; g_1, \ldots, g_t) = \mathrm{rem}_\prec(f; g_{\sigma(1)}, \ldots, g_{\sigma(t)})$$

for all permutations σ.

We can therefore write $\mathrm{rem}_\prec(f; G)$ instead of $\mathrm{rem}_\prec(f; g, \ldots, g_t)$. The following holds for polynomials which need not form a Gröbner basis.

Exercise 9.25 Let $f_1, \ldots, f_t$ be an arbitrary finite family of polynomials in $K[x_1, \ldots, x_n]$. Show that for arbitrary $f, g \in K[x_1, \ldots, x_n]$ and $c \in K$:

(a) $\mathrm{rem}_\prec(f + g; f_1, \ldots, f_t) = \mathrm{rem}_\prec(f; f_1, \ldots, f_t) + \mathrm{rem}_\prec(g; f_1, \ldots, f_t)$;

(b) $\mathrm{rem}_\prec(cf; f_1, \ldots, f_t) = c \, \mathrm{rem}_\prec(f; f_1, \ldots, f_t)$.

This implies that a Gröbner basis defines a *normal form* for the *equivalence classes*

$$f + I = \mathrm{rem}(f; G) + I.$$

Furthermore, the normal forms of the equivalence classes for I define a K-vector space.

9.4 Buchberger's Algorithm

The proof of the existence of Gröbner bases in Theorem 9.22 was not constructive. The topic of this section is an algorithm for computing Gröbner bases that dates back to the PhD thesis of Bruno Buchberger in 1965. His method is one of the most important methods in modern computer algebra.

The following finiteness statement will later provide an argument for the termination of Buchberger's algorithm.

Proposition 9.26 (Ascending Chain Condition) *Let $I_1 \subseteq I_2 \subseteq I_3 \subseteq \cdots$ be a monotonically ascending chain of ideals in $K[x_1, \ldots, x_n]$, then there exists an $N \geq 1$ with $I_N = I_{N+1} = I_{N+2} = \cdots$.*

In other words: Every ascending chain of ideals terminates.

Proof Given an ascending chain of ideals $I_1 \subseteq I_2 \subseteq I_3 \subseteq \cdots$ we study the union $I = \bigcup_{i=1}^{\infty} I_i$. Definition 9.1 gives that I is an ideal. Hilbert's Basis Theorem 9.23 shows that I has a finite set of generators $f_1, \ldots, f_t$. Every polynomial f_i is contained in an ideal I_{j_i} for a suitable $j_i \in \mathbb{N}$. For $N = \max\{j_i : 1 \leq i \leq t\}$ we have $f_1, \ldots, f_t \in I_N$ and therefore, $I_N = I_{N+1} = \cdots = I$. $\qquad\square$

A commutative ring is called *Noetherian* if the ascending chain condition holds. In the proof above we saw that the ascending chain condition follows from the fact that all ideals are finitely generated. The converse is also true: Hilbert's basis theorem and the ascending chain condition are equivalent.

As before we fix the monomial order $\prec$ for the following.

Definition 9.27 The *S-polynomial* of two non-zero polynomials f and g in $K[x_1, \ldots, x_n]$ is defined as

$$\mathrm{spol}_{\prec}(f, g) := \frac{\mathrm{lt}_{\prec}(g)}{m} f - \frac{\mathrm{lt}_{\prec}(f)}{m} g,$$

where m denotes the greatest common divisor of $\mathrm{lm}_{\prec}(f)$ and $\mathrm{lm}_{\prec}(g)$.

Buchberger's Gröbner basis algorithm uses the following characterization.

Theorem 9.28 (Buchberger's Criterion) *A finite set* $G = \{g_1, \ldots, g_t\} \subseteq K[x_1, \ldots, x_n]$ *is a Gröbner basis for* $\langle G \rangle$ *with respect to* $\prec$ *if and only if the remainder* $\mathrm{rem}_\prec(\mathrm{spol}_\prec(g_i, g_j); G)$ *vanishes for all* $i, j \in \{1, \ldots, t\}$.

Proof If G is a Gröbner basis, then we have $\mathrm{spol}(g_i, g_j) \in I$ and the remainder after Euclidean division by G is the zero polynomial.

For the reverse implication let $\mathrm{rem}(\mathrm{spol}(g_i, g_j); G) = 0$ for all i, j. A polynomial $f \in I$ has a representation

$$f = \sum_{i=1}^{t} h_i g_i \tag{9.4}$$

with polynomials $h_1, \ldots, h_t \in K[x_1, \ldots, x_n]$. We have to show that the leading term $\mathrm{lt}(f)$ is a multiple of $\mathrm{lt}(g_i)$ for some basis element $g_i \in G$. The representation (9.4) immediately gives that

$$\mathrm{lm}(f) \preceq \max\{\mathrm{lm}(h_i g_i) : 1 \leq i \leq t\} = x^\alpha$$

for an $\alpha \in \mathbb{N}^n$. Without loss of generality we can assume that $\mathrm{lm}(h_1 g_1) = x^\alpha$ and $\mathrm{lc}(g_i) = 1$ for all $i \in \{1, \ldots, t\}$. We distinguish between two cases.

Case 1: $\mathrm{lm}(f) = x^\alpha$. Here the monomial x^α is a multiple of $\mathrm{lm}(g_1)$ and we have nothing left to show.

Case 2: $\mathrm{lm}(f) \prec x^\alpha$. In this case there exists at least one other polynomial $h_i g_i$ such that $\mathrm{lt}(h_i g_i) = x^\alpha$, as otherwise it would be impossible to cancel the x^α terms through addition. Without loss of generality we can assume that $\mathrm{lm}(h_2 g_2) = x^\alpha$. Using the notation $\mathrm{lt}(h_1) = b_\beta x^\beta$ and $\mathrm{lt}(h_2) = c_\gamma x^\gamma$ we have

$$h_1 g_1 = \left(b_\beta x^\beta + \cdots\right) g_1 = b_\beta x^\beta g_1 + \left(\text{terms} \prec x^\alpha\right) \quad \text{and}$$

$$h_2 g_2 = \left(c_\gamma x^\gamma + \cdots\right) g_2 = c_\gamma x^\gamma g_2 + \left(\text{terms} \prec x^\alpha\right).$$

By construction we have that x^α is a multiple of the leading monomials of g_1 and g_2 and hence also a multiple of $x^\mu := \mathrm{lcm}(\mathrm{lm}(g_1), \mathrm{lm}(g_2))$. This yields

$$h_1 g_1 + h_2 g_2 = (b_\beta + c_\gamma) x^\beta g_1 + c_\gamma \left(x^\gamma g_2 - x^\beta g_1\right) + \left(\text{terms} \prec x^\alpha\right)$$

$$= (b_\beta + c_\gamma) x^\beta g_1 - c_\gamma x^{\alpha - \mu} \mathrm{spol}(g_1, g_2) + \left(\text{terms} \prec x^\alpha\right).$$

Our assumption implied $\mathrm{rem}(\mathrm{spol}(g_1, g_2); G) = 0$ and thus Lemma 9.14 implies that there exist polynomials $u_1, \ldots, u_t$ with

$$\mathrm{spol}(g_1, g_2) = \sum_{i=1}^{t} u_i g_i$$

Algorithm 9.4: Buchberger's algorithm

Input: finite set of polynomials $F = \{f_1, \ldots, f_t\} \subseteq K[x_1, \ldots, x_n]$
Output: Gröbner basis G for $\langle F \rangle$ with respect to $\prec$ with $F \subseteq G$

1 $G \leftarrow F$
2 **repeat**
3 $\quad$ $G' \leftarrow G$
4 $\quad$ **foreach** pair $\{p, q\} \subseteq G'$ with $p \neq q$ **do**
5 $\quad\quad$ $r \leftarrow \mathrm{rem}_\prec(\mathrm{spol}_\prec(f, g); G')$
6 $\quad\quad$ **if** $r \neq 0$ **then**
7 $\quad\quad\quad$ $G \leftarrow G \cup \{r\}$
8 **until** $G = G'$
9 **return** (G)

---

and $\mathrm{lm}(u_i g_i) \preceq \mathrm{lm}(\mathrm{spol}(g_1, g_2)) \prec x^\mu$. In particular we have $\mathrm{lm}(x^{\alpha - \mu} u_i g_i) \prec x^\alpha$ for $1 \leq i \leq t$, which implies that there exist polynomials $h'_1, \ldots, h'_t$ with

$$ f = \sum_{i=1}^{t} h'_i g_i. $$

Compared to the original representation (9.4), the number of terms $h'_i g_i$ whose leading monomial is x^α either decreases, or we have

$$ \max_\prec \big\{ \mathrm{lm}\big(h'_i g_i\big) : 1 \leq i \leq t \big\} \prec x^\alpha. $$

Therefore, after finitely many steps, the problem can be reduced to the first case. This proves the statement. $\qquad\square$

The basic idea behind the computation of a Gröbner basis of an ideal is to successively add S-polynomials to a given generating system. By Buchberger's criterion we know that we have a Gröbner basis if all of the remainders of the S-polynomials vanish when divided by the generators. We summarize the method in Algorithm 9.4.

Theorem 9.29 *Let $f_1, \ldots, f_t \in K[x_1, \ldots, x_n]$ with $\langle f_1, \ldots, f_t \rangle \neq \{0\}$. Buchberger's algorithm computes a Gröbner basis for the ideal $I = \langle f_1, \ldots, f_t \rangle$.*

Proof Every polynomial that is added to G throughout the algorithm is contained in the ideal I. Since no polynomial is ever removed from G, we retain the property $\langle G \rangle = I$ after each step. If the algorithm terminates, Buchberger's Criterion 9.28 implies that G is a Gröbner basis.

It remains to be shown that the algorithm terminates after finitely many steps. Throughout the algorithm, when $r \neq 0$ we have that $\mathrm{lt}(r) \notin \langle \mathrm{lt}(g) : g \in G \rangle$. Hence, adding r to the basis G makes the ideal $\langle \mathrm{lt}(g) : g \in G \rangle$ strictly larger. If the algorithm did not terminate, it would yield an infinitely ascending chain of ideals, contradicting Proposition 9.26. $\qquad\square$

9.5 Binomial Ideals

A polynomial of the form $x^{\alpha} - x^{\alpha'} \in K[x_1, \ldots, x_n]$ with $\alpha, \alpha' \in \mathbb{N}^n$ is called a *binomial*, and an ideal that has a generating system consisting of binomials is called a *binomial ideal*. The previously described theories are very simple in the case of binomial ideals. This will be particularly useful in Section 10.6.

Two elementary observations illustrate the uniqueness of the situation. First, we divide two binomials. For this we fix a monomial order $\prec$. If we assume for $\alpha, \alpha', \beta, \beta' \in \mathbb{N}^n$ that $x^{\alpha} \succ x^{\alpha'}$, $x^{\beta} \succ x^{\beta'}$ and that x^{β} divides x^{α}, then we get

$$x^{\alpha} - x^{\alpha'} = x^{\alpha - \beta} \cdot \left(x^{\beta} - x^{\beta'} \right) - x^{\alpha'} + x^{\alpha - \beta + \beta'}. \tag{9.5}$$

In particular,

$$\mathrm{rem}\left(x^{\alpha} - x^{\alpha'}; x^{\beta} - x^{\beta'} \right) = x^{\alpha - \beta + \beta'} - x^{\alpha'} \tag{9.6}$$

is a binomial. From this we can deduce the following.

Lemma 9.30 *Let $b_1, \ldots, b_t$ be a family of binomials. Then:*

(a) *for every monomial x^{α}, $\mathrm{rem}(x^{\alpha}; b_1, \ldots, b_t)$ is again a monomial; and*
(b) *for every binomial $x^{\alpha} - x^{\alpha'}$, $\mathrm{rem}(x^{\alpha} - x^{\alpha'}; b_1, \ldots, b_t)$ is again a binomial.*

Proof For the special case $t = 1$ we explicitly showed the second statement in (9.5). The general case $t \geq 2$ follows since we can simply iterate the computation.

The first statement follows analogously. In (9.5) we can alternatively set $\alpha' = -\infty$ with the convention that $x^{-\infty} = 0$. Then $x^{\alpha} - x^{\alpha'} = x^{\alpha}$ is a monomial and $\mathrm{rem}(x^{\alpha}; x^{\beta} - x^{\beta'}) = x^{\alpha - \beta + \beta'}$. Again, a simple iteration yields the result of the division by several polynomials. $\qquad\square$

The second observation is of similar simplicity.

Lemma 9.31 *The S-polynomial of two binomials is a binomial.*

Proof We assume $\alpha, \alpha', \beta, \beta' \in \mathbb{N}^n$ with $x^{\alpha} \succ x^{\alpha'}$ and $x^{\beta} \succ x^{\beta'}$. Furthermore, let $x^{\mu} = \gcd(x^{\alpha}, x^{\beta})$. Then we have the equation

$$\mathrm{spol}\left(x^{\alpha} - x^{\alpha'}, x^{\beta} - x^{\beta'} \right) = x^{\beta - \mu} \cdot \left(x^{\alpha} - x^{\alpha'} \right) - x^{\alpha - \mu} \cdot \left(x^{\beta} - x^{\beta'} \right)$$

$$= x^{\alpha + \beta' - \mu} - x^{\alpha' + \beta - \mu}. \qquad\square$$

When we examine the individual steps of Algorithm 9.4, the most important statement about binomial ideals follows directly from the above two lemmas.

Theorem 9.32 *Given a binomial generating system of a (necessarily binomial) ideal, Buchberger's algorithm computes a Gröbner basis consisting of binomials.*

9.6 Proving a Simple Geometric Fact Using Gröbner Bases

We now demonstrate how Gröbner bases can be employed to prove incidence statements and length relations in elementary geometry.

Theorem 9.33 *The three medians of a (non-degenerate) triangle* $\mathrm{conv}\{a, b, c\} \subseteq \mathbb{R}^2$ *intersect in a single point which we will call* s. *Each of the medians is divided by* s *in the relation* $2 : 1$.

In high school this theorem is proven directly, e.g. by setting up a system of equations that is obtained by the equations of the involved lines.

Proof Note that we can simplify our task by observing that the statement is independent of translation. That is, we can assume that the vertex a is the origin $(0, 0)$. We can choose a second point, say b, as $(1, 0)$ since the statement is independent of rotation and scaling. We denote the coordinates of the third point by $c = (x, y)$.

We use the notation from Fig. 9.3. The three midpoints of the sides have coordinates

$$p = \left(\frac{x + 1}{2}, \frac{y}{2}\right), \qquad q = \left(\frac{x}{2}, \frac{y}{2}\right), \qquad r = \left(\frac{1}{2}, 0\right).$$

Let $s := (u, v)$ be the intersection of $\mathrm{aff}(a, p)$ and $\mathrm{aff}(b, q)$. The fact that s lies on $\mathrm{aff}(a, p)$ is (by comparing the slope of the lines $\mathrm{aff}(a, s)$ and $\mathrm{aff}(a, p)$) equivalent to

$$f_1 := uy - v(x + 1) = 0.$$

Analogously, the relation $s \in \mathrm{aff}(b, q)$ is equivalent to

$$f_2 := (u - 1)y - v(x - 2) = 0.$$

s lies on $\mathrm{aff}(c, r)$ if and only if

$$g_1 := -2(u - x)y - (v - y)(1 - 2x) = -2uy - (v - y) + 2vx = 0.$$

The point s divides the medians in a $2 : 1$ relation if and only if the following three equations hold:

$$(u, v) = s - a = 2(p - s) = (x + 1 - 2u, y - 2v),$$

$$(u - 1, v) = s - b = 2(q - s) = (x - 2u, y - 2v),$$

$$(u - x, v - y) = s - c = 2(r - s) = (2u - 1, 2v).$$

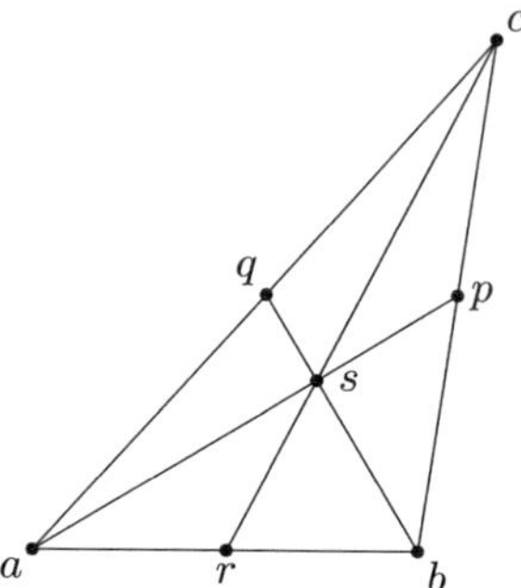

Fig. 9.3 The medians of a triangle meet in a common point s, which is in fact the *center of mass*

This reduces to

$$g_2 := 3u - x - 1 = 0,$$

$$g_3 := 3v - y = 0.$$

We have to respect the condition that our triangle $\mathrm{conv}\{a, b, c\}$ is not degenerate, i.e., $y \neq 0$. This can be expressed by an equation if we introduce another variable z:

$$f_3 := yz - 1 = 0.$$

Now we want to show that

$$f_1 = f_2 = f_3 = 0 \quad \Longrightarrow \quad g_1 = g_2 = g_3 = 0$$

or, in other words, that $\mathrm{V}(f_1, f_2, f_3) \subseteq \mathrm{V}(g_1, g_2, g_3)$. Our proof is complete if we can show the stronger statement

$$g_1, g_2, g_3 \in \langle f_1, f_2, f_3 \rangle.$$

We compute a Gröbner basis of the ideal $I := \langle f_1, f_2, f_3 \rangle \subseteq \mathbb{R}[u, v, x, y, z]$ for, say, the graded reverse lexicographic order $\prec_{\mathrm{grevlex}}$. Using Buchberger's Criterion 9.28 we can verify that

$$G = \{3v - y, 3u - x - 1, yz - 1\}$$

is a $\prec_{\mathrm{grevlex}}$-Gröbner basis of I. Dividing out three candidates g_1, g_2, g_3 by G yields

$$\mathrm{rem}(g_1; G) = \mathrm{rem}(g_2; G) = \mathrm{rem}(g_3; G) = 0,$$

i.e., $g_1, g_2, g_3 \in I$. $\qquad\square$

Observe that we didn't assume x, y, u and v to be real numbers. The proof is therefore also valid over $\mathbb{C}$.

9.7 Exercises

Exercise 9.34 Show that, given two univariate polynomials $f, g \in K[x] \setminus \{0\}$, there exist polynomials $a, b \in K[x]$ such that

$$\gcd(f, g) = af + bg.$$

To do so, analyze the Euclidean Algorithm 9.1 and modify it in such a way that the polynomials a and b are computed.

The method described in Exercise 9.34 is called the *extended Euclidean algorithm*.

Exercise 9.35 Let $G = \{g_1, \ldots, g_t\}$ be a Gröbner basis of an ideal $I \subseteq K[x_1, \ldots, x_n]$ with respect to the monomial order $\prec$ and let f, g be polynomials whose difference $f - g$ lies in I. Show that $g = \mathrm{rem}_{\prec}(f; G)$ if and only if no term of g is divisible by one of the leading monomials of $\mathrm{lt}_{\prec}(g_1), \ldots, \mathrm{lt}_{\prec}(g_t)$.

For a Gröbner basis G of an ideal I, we have that every superset G' of G with $G' \subseteq I$ is a Gröbner basis of I. This leads to the question if a given Gröbner basis can have superfluous elements.

Definition 9.36 A Gröbner basis G of an ideal I is called *reduced* if for all $g \in G$:

(a) The leading coefficient is normalized: $\mathrm{lc}_{\prec}(g) = 1$.
(b) No monomial of g lies in $\mathrm{in}_{\prec}(G \setminus \{p\})$.

Exercise 9.37 Show that every non-zero ideal has a unique reduced Gröbner basis for the monomial order $\prec$.

9.8 Remarks

The structure of our presentation is based on the beautiful and comprehensive introduction to the theory of Gröbner bases by Cox, Little and O'Shea [28]. Another text worth reading is the monograph of Adams and Loustaunau [1]. The example of the geometric proof was taken from zur Gathen and Gerhard [97].

Gröbner bases were introduced in the 1960s by Hironaka [60, 61] (who called them "standard bases") and independently by Buchberger in his dissertation [17] in 1965. The term "Gröbner basis" was established by Buchberger in honor of his PhD advisor Wolfgang Gröbner. The exact origin of the "S" in the term "S-polynomial" is not clear. It is sometimes interpreted as "subtraction" or "syzygy".

The statement of the Gordan–Dickson Lemma 9.20 was (re)discovered several times. Its first explicit appearances are usually credited to the German mathematician Paul Gordan [50] and to the American mathematician Leonard Eugene Dickson [35].

If the coefficients of two polynomials f and g are rational numbers, then the computation of the greatest common divisor via the Euclidean algorithm is performed in polynomial time. As stated in Appendix C, polynomial time performance refers to the total length of the input coded as a series of bits. In contrast to this, the ideal membership problem, as well as the problem of computing a Gröbner basis, are intrinsically difficult problems. Mayr and Meyer [76] showed that, with respect to complexity theory, every problem that can be solved with an exponentially large memory can be reduced to an ideal membership problem. Since an exponentially large memory is sufficient, we have that the ideal membership problem is EXPSPACE-*complete*. EXPSPACE-complete problems are significantly more difficult than NP-complete problems: All known algorithms for EXPSPACE-complete problems have at least double-exponential worst-case run-time.

From a practical viewpoint, Buchberger's algorithm can be more efficient in several ways, e.g. by avoiding the computation of superfluous S-polynomials (besides the aforementioned books, see also *Using Algebraic Geometry* by Cox, Little and O'Shea [29] as well as the book by Becker and Weispfenning [11]).

Algorithmic concepts which occur in the solution of problems in the field of *real algebraic geometry* include a variety of methods which are not mentioned in this book. For an overview we refer to the monograph by Basu, Pollack and Roy [10]. Additionally, over the real numbers the question of how to deal with systems of polynomial *inequalities* arises. This leads to *semi-algebraic geometry*. For this, Collins developed an important approach called the *cylindric algebraic decomposition* [25] (for quantifier elimination over real-closed fields). This method is implemented in QEPCAD [65].

Chapter 10
Solving Systems of Polynomial Equations Using Gröbner Bases

The focus of this chapter is on a general method for solving systems of polynomial equations via Gröbner bases. We will first briefly present how the computer algebra systems `Maple` and `Singular` can be used to compute Gröbner bases and solve systems of polynomial equations. We illustrate the methods discussed in later sections of this chapter with an analysis of several examples using `Maple` and `Singular`. A short introduction to these programs can be found in Appendix D.

When solving systems of polynomial equations, we must also determine the conditions a system needs to satisfy in order to have solutions. This leads to Hilbert's Nullstellensatz, which we prove in Section 10.4.

Finally, we sketch in Section 10.6, a possibly unexpected, application of elimination theory to integer linear programs.

10.1 Gröbner Bases Using `Maple` and `Singular`

Standard computer algebra systems provide several methods for the computation of Gröbner bases. We begin by illustrating some computations with the commercial mathematical software system `Maple`. Our main goal is to demonstrate the effective availability of the specific algorithms within the computer algebra packages, and to motivate the reader to use them. We do not detail the slight variations in the syntax for specific commands of different software packages.

To be able to use algorithms for the computation of Gröbner bases in `Maple`, we first have to load the package `Groebner`. If a command line ends with a colon instead of a semicolon, `Maple` suppresses the output.

```
> with(Groebner):
```

We now compute a Gröbner basis of the ideal $I = \langle xy + 1, yz + 1 \rangle$ in $\mathbb{C}[x, y, z]$ with respect to the lexicographic monomial order (which is called `plex` in `Maple`).

```
> G:=[x*y+1,y*z+1]:
> Basis(G,plex(x,y,z));
```

M. Joswig, T. Theobald, *Polyhedral and Algebraic Methods in Computational Geometry*, Universitext, DOI 10.1007/978-1-4471-4817-3_10,
© Springer-Verlag London 2013

The output is

```
[y z + 1, -z + x].
```

That is, the polynomials $yz + 1$ and $x - z$ form a Gröbner basis of the ideal I.

The computation of a Gröbner basis with respect to the graded reverse lexicographic order (called `tdeg` in `Maple`) via

```
> Basis(G,tdeg(x,y,z));
```

yields the output

```
[-z + x, y z + 1].
```

In this case, both monomial orders produce the same Gröbner basis.

The free software package `Singular` (which is also part of the `Sage` software system) is much more specialized for methods based on Gröbner bases than `Maple`. Furthermore, the number of methods available in `Singular` is larger. To begin, we need to specify the base ring in `Singular`.

```
> ring R = 0, (x,y,z), lp;
```

declares that we are working in the polynomial ring $\mathbb{Q}[x, y, z]$. The coefficient field $\mathbb{Q}$ is the prime field of characteristic 0, which is why it is written as "0" in `Singular`; similarly, a prime number in this position would declare the corresponding finite (prime) field. The parameter `lp` causes all of the following computations to use the lexicographic order. The graded reverse lexicographic order can be used by writing `dp`.

To compute the Gröbner basis of the above example in `Singular`, we define the ideal $I = \langle xy + 1, yz + 1 \rangle$ via its two generators:

```
> ideal I = x*y+1, y*z+1;
```

The computation of the Gröbner basis with respect to the lexicographic order via

```
> groebner(I);
```

yields the output of the polynomials (with numeric labels)

```
_[1]=yz+1
_[2]=x-z.
```

The input

```
> quit;
```

causes `Singular` to politely exit saying

```
Auf Wiedersehen.
```

10.2 Elimination of Unknowns

As announced in Section 9.1, we will now study *elimination ideals*

$$I_k := I \cap K[x_{k+1}, \ldots, x_n], \quad \text{for } 0 \leq k < n,$$

of an ideal I in $K[x_1, \ldots, x_n]$. From Exercise 9.4, we know that I_k is an ideal in $K[x_{k+1}, \ldots, x_n]$. The lexicographic order has a special property with regard to elimination ideals.

Theorem 10.1 *Let I be an ideal in $K[x_1, \ldots, x_n]$ and G be a Gröbner basis of I with respect to the lexicographic order $x_1 \succ_{\mathrm{lex}} \cdots \succ_{\mathrm{lex}} x_n$. Then*

$$G_k := G \cap K[x_{k+1}, \ldots, x_n]$$

is a Gröbner basis for the k-th elimination ideal I_k with $0 \le k < n$.

Proof Let $k \in \{0, \ldots, n-1\}$ and let $G = \{g_1, \ldots, g_t\}$. Without loss of generality we can assume $G_k = \{g_1, \ldots, g_s\}$ for an $s \le t$. First, we show that G_k generates the ideal I_k. Since $G_k \subseteq I_k$, it suffices to show that every polynomial $f \in I_k$ can be written as a linear combination of $g_1, \ldots, g_s$ with coefficients in $K[x_{k+1}, \ldots, x_n]$.

Since $f \in I$, by Lemma 9.14, dividing the polynomial f by the ordered series of the Gröbner basis G gives a representation

$$f = h_1 g_1 + \cdots + h_t g_t$$

with $h_1, \ldots, h_t \in K[x_1, \ldots, x_n]$ and

$$\mathrm{lm}(f) \succeq_{\mathrm{lex}} \mathrm{lm}(h_i g_i).$$

By construction, at least one of the unknowns x_i with $i \le k$ occurs in every polynomial $g_{s+1}, \ldots, g_t$. Thus, $\mathrm{lm}(g_i) \succ_{\mathrm{lex}} \mathrm{lm}(f)$ for $i \in \{s+1, \ldots, t\}$ and $h_{s+1}, \ldots, h_t = 0$. With this we obtain the desired representation

$$f = h_1 g_1 + \cdots + h_s g_s.$$

To prove that $\{g_1, \ldots, g_s\}$ is a Gröbner basis, we show that it satisfies Buchberger's Criterion 9.28. Therefore, we need that each S-polynomial $\mathrm{spol}(g_i, g_j)$ for $1 \le i \ne j \le s$ has remainder zero after division by $g_1, \ldots, g_s$. Since $\mathrm{spol}(g_i, g_j) \in I_k$, this follows from the first part of the proof. $\square$

The last elimination ideal $I_{n-1} \subseteq K[x_n]$ is univariate. Hence, the corresponding reduced Gröbner basis G_{n-1} consists of only one element, which is called the *eliminant* of I, or we have $I_{n-1} = \{0\}$.

Example 10.2 We return to Example 8.1 and perform the calculations with `Maple`.

```
> with(Groebner):
> f := x^2+y^2-x*y-x-y-1:
> g := 2*x^2-4*y^2-x*y-2*x-2*y-1:
> G := Basis([f,g],plex(x,y));
```

We obtain

```
                 2       3      4              2       3
  G := [1 + y - 12 y  - 7 y  + 31 y , -1 + 6 y + 7 y  - 31 y  + x]
```

Thus, the first (and last) elimination ideal I_1 of $I = \langle f, g \rangle$ is generated by the eliminant $p := 31y^4 - 7y^3 - 12y^2 + y + 1$.

For every point $(\xi, \eta) \in V(I)$, the number η is a root of p. Since p is of degree 4, it would be possible to compute explicit representations of its zeros in terms of radicals. We will not do this here, but instead choose a strategy that also works for higher degrees: We only compute numerical approximations of the zeros; see the discussion in Section 8.1.

```
> p := G[1];
> fsolve(p=0,y);
```

This yields

```
   -0.4416023314,  -0.3223146983,  0.3597572748,  0.6299662065.
```

The numerical values of the x-coordinates, and with this the intersection points of the two conic sections, can be obtained with the following:

```
> eta := [fsolve(p=0,y)]:
> seq([ fsolve(subs(y=eta[i],G[2]=0),x), eta[i] ], i=1..4);
```

Compare the output

```
   [-0.3851331910, -0.4416023314],   [1.168669669, -0.3223146983],
   [-0.6211082730, 0.3597572748],    [2.192410502, 0.6299662065]
```

with Fig. 8.2.

We saw in the last example how to obtain the variety from the roots of the eliminant. Two questions remain to be answered. First, does every root of an eliminant lead to a point of the variety? This will be discussed in the following section. Secondly, under which conditions does an eliminant exist?

Example 10.3 As an example, take the ideal $\langle xy \rangle \subseteq K[x, y]$ with the Gröbner basis $\{xy\}$ (for any monomial order). Here, we see that an eliminant does not always exist.

We discuss this phenomenon in more detail in Section 10.5. First, we will establish a connection between elimination ideals and resultants.

Proposition 10.4 *Let* $f, g \in K[x_1, \ldots, x_n]$ *with positive degree in* x_1. *Then there exist polynomials* $a, b \in K[x_1, \ldots, x_n]$ *such that*

$$af + bg = \mathrm{Res}_{x_1}(f, g).$$

In particular, $\mathrm{Res}_{x_1}(f, g)$ *is contained in the first elimination ideal of* $\langle f, g \rangle$.

Proof Let $f, g \in K[x_1, \ldots, x_n]$ be given in the form

$$f = a_l x_1^l + \cdots + a_1 x_1 + a_0,$$
$$g = b_m x_1^m + \cdots + b_1 x_1 + b_0$$

with $a_i, b_j \in K[x_2, \ldots, x_n]$ and $a_l, b_m \neq 0$. Then the Sylvester matrix of f and g is

$$\left.\begin{pmatrix} a_l & a_{l-1} & \cdots & a_0 & & & \\ & \ddots & \ddots & & \ddots & & \\ & & a_l & a_{l-1} & \cdots & a_0 \\ b_m & b_{m-1} & \cdots & b_0 & & \\ & \ddots & \ddots & & \ddots & \\ & & b_m & b_{m-1} & \cdots & b_0 \end{pmatrix}\begin{matrix} \left.\vphantom{\begin{matrix}a\\b\\c\end{matrix}}\right\} m \text{ rows,} \\ \\ \left.\vphantom{\begin{matrix}a\\b\\c\end{matrix}}\right\} l \text{ rows.} \end{matrix}\right. \tag{10.1}$$

We want to compute the resultant of (10.1) by modifying the matrix via the following elementary column operation. Adding the i-th column multiplied by x_1^{l+m-i} to the last column gives

$$M = \begin{pmatrix} a_l & a_{l-1} & \cdots & a_0 & & & x_1^{m-1} f \\ & a_l & a_{l-1} & \cdots & a_0 & & x_1^{m-2} f \\ & & \ddots & \ddots & & \ddots & \vdots \\ & & & a_l & \cdots & a_1 & f \\ b_m & b_{m-1} & \cdots & b_0 & & & x_1^{l-1} g \\ & b_m & b_{m-1} & \cdots & b_0 & & x_1^{l-2} g \\ & & \ddots & \ddots & & \ddots & \vdots \\ & & & b_m & \cdots & b_1 & g \end{pmatrix}. \tag{10.2}$$

Expanding along the last column leads to

$$\mathrm{Res}_{x_1}(f, g) = \det M = x_1^{m-1} f \cdot p_1 + \cdots + f \cdot p_m + x_1^{l-1} g \cdot q_1 + \cdots + g \cdot q_l$$

with polynomials $p_i, q_j \in K[x_2, \ldots, x_n]$. Reordering gives the desired linear combination

$$\mathrm{Res}_{x_1}(f, g) = \left(p_1 x_1^{m-1} + \cdots + p_m\right) \cdot f + \left(q_1 x_1^{l-1} + \cdots + q_l\right) \cdot g = af + bg$$

with $a = p_1 x_1^{m-1} + \cdots + p_m$ and $s = q_1 x_1^{l-1} + \cdots + q_l$. $\qquad\square$

To explicitly construct a (non-zero) polynomial in an elimination ideal, in the case where we have an ideal $\langle f_1, \ldots, f_t \rangle$ which is generated by more than two polynomials, we use the parametric version

$$\mathrm{Res}_{x_1}(f_1, \lambda_2 f_2 + \cdots + \lambda_t f_t) \in K[x_2, \ldots, x_n, \lambda_2, \ldots, \lambda_t]$$

with parameters $\lambda_2, \ldots, \lambda_t$. The distinction between *unknowns* and *parameters* is from a formal viewpoint arbitrary. However, it hints that we will later deduce statements about polynomials in certain unknowns with respect to special parameters. Compare this with the discussion of multivariate resultants prior to Corollary 8.10.

Lemma 10.5 *Assume* $I = \langle f_1, \ldots, f_t \rangle \subseteq K[x_1, \ldots, x_n]$, *and* $\mathrm{Res}_{x_1}(f_1, \lambda_2 f_2 + \cdots + \lambda_t f_t)$ *has a representation* $\sum_\alpha h_\alpha \lambda^\alpha$ *with polynomials* $h_\alpha \in K[x_2, \ldots, x_n]$. *Then each of the polynomials* h_α *is contained in the first elimination ideal* $I_1 = I \cap K[x_2, \ldots, x_n]$.

Proof Since every polynomial h_α depends only on the unknowns $x_2, \ldots, x_n$, it suffices to show that every h_α is contained in I. By Proposition 10.4, $\mathrm{Res}_{x_1}(f_1, \lambda_2 f_2 + \cdots + \lambda_t f_t)$ has a representation of the form

$$\mathrm{Res}_{x_1}(f_1, \lambda_2 f_2 + \cdots + \lambda_t f_t) = a f_1 + b(\lambda_2 f_2 + \cdots + \lambda_t f_t)$$

with polynomials $a, b \in K[x_2, \ldots, x_n, \lambda_2, \ldots, \lambda_t]$. We write a and b as polynomials in terms of the parameters $\lambda_2, \ldots, \lambda_t$,

$$a = \sum_\alpha a_\alpha \lambda^\alpha, \qquad b = \sum_\alpha b_\alpha \lambda^\alpha,$$

with coefficients $a_\alpha, b_\alpha \in K[x_2, \ldots, x_n]$ and obtain

$$\sum_\alpha h_\alpha \lambda^\alpha = \left(\sum_\alpha a_\alpha \lambda^\alpha \right) f_1 + \left(\sum_\alpha b_\alpha \lambda^\alpha \right) \left(\sum_{i=2}^{t} \lambda_i f_i \right)$$

$$= \sum_\alpha \left(a_\alpha f_1 + \sum_{i=2}^{t} b_{\alpha + e^{(i)}} f_i \right) \lambda^\alpha,$$

where $e^{(i)}$, for $2 \leq i \leq m$, denotes the i-th unit vector in the coordinates $(\lambda_2, \ldots, \lambda_t)$. Comparing coefficients gives a representation of the polynomials h_α in terms of the generators $f_1, \ldots, f_t$. $\qquad\square$

10.3 Continuation of Partial Solutions

Up until now, we have developed the theory of Gröbner bases over an arbitrary field. It should be mentioned, however, that the process of solving arbitrary systems of polynomial equations must utilize properties of the base field. Here, we concentrate on the complex numbers as an algebraically closed field of characteristic 0.

From now on, we study an ideal $I \subseteq \mathbb{C}[x_1, \ldots, x_n]$ and the corresponding k-th elimination ideal $I_k = I \cap \mathbb{C}[x_{k+1}, \ldots, x_n]$ for a $k \in \{1, \ldots, n-1\}$. Having proved Theorem 10.1, the most important remaining open question is which conditions are necessary in order for a partial solution $(\xi_{k+1}, \ldots, \xi_n) \in \mathrm{V}(I_k)$ to be extendable to a solution $(\xi_1, \ldots, \xi_n) \in V(I)$. By induction, it suffices to determine necessary conditions for the last extension step.

Theorem 10.6 *Let $f_1, \ldots, f_t \in \mathbb{C}[x_1, \ldots, x_n]$, $I = \langle f_1, \ldots, f_t \rangle$ and let I_1 be the first elimination ideal of I. Furthermore, for $i \in \{1, \ldots, t\}$ let*

$$f_i = g_{i,d_i} x_1^{d_i} + \cdots + g_{i,0}$$

with $\deg_{x_1} f_i = d_i$ and $g_{i,j} \in \mathbb{C}[x_2, \ldots, x_n]$. Then, for all $(\xi_2, \ldots, \xi_n) \in V(I_1) \setminus V(g_{1,d_1}, \ldots, g_{t,d_t})$ there exists a $\xi_1 \in \mathbb{C}$ with $(\xi_1, \ldots, \xi_n) \in V(I)$.

Proof Only the leading coefficient polynomials of f_i in the unknown x_1 are important in the following. Thus, we write $g_i := g_{i,d_i}$.

Let $\xi = (\xi_2, \ldots, \xi_n) \in V(I_1) \setminus V(g_1, \ldots, g_t)$. Without loss of generality, we can assume $g_1(\xi) \neq 0$. The resultant $\mathrm{Res}(f_1, \lambda_2 f_2 + \cdots + \lambda_t f_t)$ with parameters $\lambda_2, \ldots, \lambda_t$ has a representation of the form

$$\mathrm{Res}_{x_1}(f_1, \lambda_2 f_2 + \cdots + \lambda_t f_t) = \sum_{\alpha} h_\alpha \lambda^\alpha \tag{10.3}$$

with polynomials $h_\alpha \in \mathbb{C}[x_2, \ldots, x_n]$. By Lemma 10.5, every polynomial h_α is contained in the elimination ideal I_1. Since $\xi \in V(I_1)$, the resultant $\mathrm{Res}_{x_1}(f_1, \lambda_2 f_2 + \cdots + \lambda_t f_t)$ vanishes at the point ξ, i.e.,

$$\mathrm{Res}_{x_1}(f_1, \lambda_2 f_2 + \cdots + \lambda_t f_t)|_{(x_2, \ldots, x_n) = \xi} = 0. \tag{10.4}$$

We now take advantage of the vanishing of this resultant at ξ. The only problem is that the x_1-degree of the polynomial $\lambda_2 f_2 + \cdots + \lambda_t f_t$ may decrease when evaluated at ξ. To avoid this, we change the basis of I.

For an arbitrary $N \in \mathbb{N}$, the polynomials $f_1, \ldots, f_{t-1}, f_t + x_1^N f_1$ generate the ideal I. So we can choose N large enough such that

$$\deg_{x_1} f_t + x_1^N f_1 > \deg_{x_1} f_i, \quad \text{for all } i \in \{2, \ldots, t\}.$$

Since $g_1(\xi) \neq 0$ and $\deg_{x_1} f_t + x_1^N f_1 > \deg_{x_1} f_t$, we know that $g_t(\xi) \neq 0$ still holds after the change of basis.

Since $g_1(\xi) \neq 0$, $g_t(\xi) \neq 0$ and $\deg_{x_1} g_t > \deg_{x_1} g_2, \ldots, \deg_{x_1} g_{t-1}$, building the resultant and substituting (in the last $n - 1$ unknowns) can be switched. Thus we have, by (10.4),

$$\mathrm{Res}_{x_1}\big(f_1(x_1, \xi), \lambda_2 f_2(x_1, \xi) + \cdots + \lambda_t f_t(x_1, \xi)\big)$$

$$= \mathrm{Res}_{x_1}(f_1, \lambda_2 f_2 + \cdots + \lambda_t f_t)|_{(x_2, \ldots, x_n) = \xi} = 0.$$

The univariate polynomials $f_1(x_1, \xi), \ldots, f_t(x_1, \xi)$ have a common factor of positive degree by Corollary 8.10. Since $\mathbb{C}$ is algebraically closed, there exists a common root ξ_1, and $(\xi_1, \ldots, \xi_n) = (\xi_1, \xi) \in V(I)$. $\square$

Example 10.7 We again study Example 10.2. Here,

$$f = x^2 + y^2 - xy - x - y - 1 = 1 \cdot x^2 - (y+1) \cdot x + \left(y^2 - y - 1\right) \quad \text{and}$$

$$g = 2x^2 - 4y^2 - xy - 2x - 2y - 1 = 2 \cdot x^2 - (y+2) \cdot x - \left(4y^2 + 2y + 1\right).$$

For $I = \langle f, g \rangle$,

$$\left\{x - 31y^3 + 7y^2 + 6y - 1, \ 31y^4 - 7y^3 - 12y^2 + y + 1\right\}$$

is a $\prec_{\mathrm{lex}}$-Gröbner basis of I.

We saw in Example 10.2 that all roots of the eliminant $31y^4 - 7y^3 - 12y^2 + y + 1$ can be extended to points in $V(I)$. We know this is true by Theorem 10.6 and the fact that (with respect to x) the leading coefficient polynomials 1 and 2 are constant, and therefore the set of common zeros of the leading coefficient polynomials is empty.

It remains to be determined exactly which conditions guarantee the existence of an eliminant. To do this, in the next section we study a fundamental result of commutative algebra.

10.4 The Nullstellensatz

If a non-zero constant polynomial is contained in I, then we clearly have $V(I) = \emptyset$. The weak form of Hilbert's Nullstellensatz states that, over the complex numbers, the reverse implication is also true.

Theorem 10.8 (Nullstellensatz, weak form) *Let I be an ideal in $\mathbb{C}[x_1, \dots, x_n]$ such that* $V(I) = \emptyset$, *then* $1 \in I$.

The property $1 \in I$ is clearly equivalent to $I = \mathbb{C}[x_1, \dots, x_n]$. An ideal I with $I \subsetneq \mathbb{C}[x_1, \dots, x_n]$ is called a *proper ideal* of $\mathbb{C}[x_1, \dots, x_n]$.

Similar to the characterization of the feasibility of linear programming problems by Farkas' lemma, see Exercise 4.26, the Nullstellensatz characterizes the solvability of a system of polynomial equations.

Example 10.9 The polynomials $f = x^2$ and $g = 1 - xy$ do not have a common root in $\mathbb{C}^2$. In order to prove this, it obviously suffices to provide a pair of polynomials (a, b) with

$$1 = af + bg. \tag{10.5}$$

Independent of how difficult it is to determine such a and b, once we know them it is rather easy to verify the identity (10.5). Therefore we call the pair (a, b) a *certificate* for the non-existence of common roots of f and g.

For our example of f and g, a pair of polynomials (a, b) satisfying (10.5) is $a = y^2$ and $b = 1 + xy$. The Nullstellensatz guarantees the existence of such polynomials, without explicitly stating them.

We describe here the proof of Hilbert's Nullstellensatz by Arrondo [5], which is formulated in relatively elementary terms. The following lemma will be employed to transform the polynomials into a form which simplifies the analysis.

Lemma 10.10 (Noetherian Normalization Lemma)   *Let $n \geq 2$ and $f \in \mathbb{C}[x_1, \ldots, x_n]$ be a non-constant polynomial of total degree d. Then there exist complex numbers $\lambda_2, \ldots, \lambda_n$ such that the monomial x_1^d has a non-zero coefficient in the polynomial*

$$f(x_1, x_2 + \lambda_2 x_1, \ldots, x_n + \lambda_n x_1). \tag{10.6}$$

Proof From tdeg $f = d$ it follows that $\deg_{x_1} f(x_1, x_2 + \lambda_2 x_1, \ldots, x_n + \lambda_n x_1) = d$. If f_d denotes the homogeneous component of f of degree d, then we can represent the coefficient of x_1^d in (10.6) as $f_d(1, \lambda_2, \ldots, \lambda_n)$. Since the polynomial $f_d(1, x_2, \ldots, x_n)$ is not the zero polynomial, there exists a point $(\lambda_2, \ldots, \lambda_n) \in \mathbb{C}^{n-1}$ at which it does not vanish (see Exercise 10.30). $\qquad\square$

We are now able to prove the Nullstellensatz in its weak form.

Proof of Theorem 10.8 We prove the contrapositive: For every proper ideal $I \subsetneq \mathbb{C}[x_1, \ldots, x_n]$, there exists a $\xi \in \mathbb{C}^n$ such that $f(\xi_1, \ldots, \xi_n) = 0$ for all $f \in I$.

Without loss of generality, let $I \neq \{0\}$. The statement is clear for $n = 1$, since $\mathbb{C}[x_1]$ is a Euclidean ring, i.e., every ideal I is generated by a non-constant polynomial. By the fundamental theorem of algebra, every such generator of I has a root.

We handle the case $n \geq 2$ inductively. By Lemma 10.10 we can assume that I contains a normalized polynomial g in the unknown x_1.

$$I_1 = I \cap \mathbb{C}[x_2, \ldots, x_n] \subseteq \mathbb{C}[x_2, \ldots, x_n]$$

is the first elimination ideal of I. Since $1 \notin I$, I_1 is a proper ideal. By the inductive hypothesis, there exists a point $(\xi_2, \ldots, \xi_n) \in \mathbb{C}^{n-1}$ at which all polynomials from I_1 vanish.

The crucial step is to show that the set

$$J = \left\{ f(x_1, \xi_2, \ldots, \xi_n) : f \in I \right\}$$

is a proper ideal of $\mathbb{C}[x_1]$.

By the distributive laws, it is clear that J is an ideal. To complete this step, we use an indirect approach. Assume that $1 \in J$, i.e., there exists a $g(x_1, \xi_2, \ldots, \xi_n) = 1$. If g has x_1-degree d, there exists a representation of the form $g = \sum_{i=0}^{d} g_i x_1^i$ with $g_0, \ldots, g_d \in \mathbb{C}[x_2, \ldots, x_n]$, $g_0(\xi_2, \ldots, \xi_n) = 1$, and $g_i(\xi_2, \ldots, \xi_n) = 0$ for $1 \leq i \leq d$.

The x_1-normalized polynomial f with $\deg_{x_1} f = e$, which we obtained above, can be written in the form $f = x_1^e + \sum_{i=0}^{e-1} f_i x_1^i$ with $f_i \in \mathbb{C}[x_2, \ldots, x_n]$. By Proposition 10.4, we know that the resultant $\mathrm{Res}_{x_1}(f, g)$ is contained in the elimination ideal I_1. Since

$$\mathrm{Res}_{x_1}(f,g) = \det \begin{pmatrix} 1 & f_{e-1} & \cdots & f_0 & & & \\ & \ddots & \ddots & & \ddots & & \\ & & 1 & f_{e-1} & \cdots & f_0 \\ g_d & g_{d-1} & \cdots & g_0 & & \\ & \ddots & \ddots & & \ddots & \\ & & g_d & g_{d-1} & \cdots & g_0 \end{pmatrix} \left.\begin{matrix} \\ \\ \\ \end{matrix}\right\} d \text{ rows,} \atop \left.\begin{matrix} \\ \\ \end{matrix}\right\} e \text{ rows}$$

the resultant $\mathrm{Res}_{x_1}(f,g)$ at the point $(\xi_2,\ldots,\xi_n)$ can be evaluated as the determinant of an upper triangular matrix, with all entries on the main diagonal equal to 1. Therefore, $\mathrm{Res}_{x_1}(f,g)$ is 1 at $(\xi_2,\ldots,\xi_n)$, which contradicts $\mathrm{Res}_{x_1}(f,g) \in I_1$ and $(\xi_2,\ldots,\xi_n) \in V(I_1)$. This proves the statement $J \subsetneq \mathbb{C}[x_1]$.

Therefore, the ideal J is generated by a polynomial $h \in \mathbb{C}[x_1]$ of positive degree, or by $h = 0$. In both cases, h has at least one root $\xi_1 \in \mathbb{C}$. Thus, every polynomial of I vanishes at $(\xi_1,\ldots,\xi_n)$. $\qquad\square$

The following strong form of the Nullstellensatz can be obtained from the weak form. We present here the key to the proof which is often referred to as the "trick of Rabinowitsch".

Theorem 10.11 (Nullstellensatz, strong form) *If I is an ideal in $\mathbb{C}[x_1,\ldots,x_n]$, and $f \in \mathbb{C}[x_1,\ldots,x_n]$ a polynomial which vanishes at all points of $V(I)$, then there exists a natural number $s \geq 1$ such that $f^s \in I$.*

Proof We assume that $g_1,\ldots,g_t$ generate the ideal I.

If $f = 0$, then nothing remains to be shown. For $f \neq 0$ we study the polynomials

$$g_1,\ldots,g_t, 1-yf \in \mathbb{C}[x_1,\ldots,x_n,y].$$

These do not have a common zero, since for all $\xi = (\xi_1,\ldots,\xi_n) \in V(g_1,\ldots,g_t)$ we have that $1-yf$ has the value

$$1 - \eta f(\xi_1,\ldots,\xi_n) = 1 - \eta \cdot 0 = 1$$

at the point $(\xi_1,\ldots,\xi_n,\eta)$. The weak form of Hilbert's Nullstellensatz yields $h_1,\ldots,h_{t+1} \in \mathbb{C}[x_1,\ldots,x_n,y]$ such that

$$h_1 g_1 + \cdots + h_t g_t + h_{t+1}(1-yf) = 1. \tag{10.7}$$

If we perform our computations in the quotient field $\mathbb{C}(x_1,\ldots,x_n,y)$ of rational functions, we can substitute the rational function $1/f$ for the unknown y in (10.7). Thus, there exist $h'_1,\ldots,h'_t \in K[x_1,\ldots,x_n]$ and $s_1,\ldots,s_t \in \mathbb{N}$ such that

$$\frac{h'_1}{f^{s_1}} g_1 + \cdots + \frac{h'_m}{f^{s_t}} g_t = 1.$$

Choosing $s = \max\{s_1,\ldots,s_t\}$ and multiplying by f^s yields the statement. $\qquad\square$

Example 10.12 The Nullstellensatz implies for an arbitrary monomial order that the uniquely determined reduced Gröbner basis of an ideal I consists of the constant polynomial 1 if and only if $V(I) = \emptyset$.

We use `Maple` to verify this statement for the polynomials $f = x^2$, $g = 1 - xy$ from Example 10.9:

```
> with(Groebner):
> Basis([x^2,1-xy],plex(x,y));
  [1]
```

Note that Study's Lemma 8.12 is a special case of the Nullstellensatz. Let f be a non-constant, irreducible polynomial such that $V(f) \subseteq V(g)$, then the Nullstellensatz implies that $g^k \in \langle f \rangle$ for some $k \geq 1$, and hence that f divides g.

Definition 10.13 For an ideal I of an arbitrary ring R the set

$$\mathrm{rad}(I) := \left\{ a \in R : \text{there exists a } s \geq 1 \text{ such that } a^s \in I \right\}$$

is called the *radical* of I in R.

Example 10.14 Consider the ideal $I = \langle x^2, xy, y^2 \rangle \subseteq \mathbb{C}[x, y]$. Clearly, we have $V(I) = \{(0, 0)\}$. The polynomial x vanishes at the unique zero of I, so that Hilbert's Nullstellensatz implies that a suitable power of x must be in I. The polynomial x^2 was explicitly stated as a generator of I. With a little more work, we see that $\langle x, y \rangle$ is the radical of I.

10.5 Solving Systems of Polynomial Equations

Now, we will combine the main statements from the theory of Gröbner bases and show how to use them to systematically solve systems of polynomial equations.

Theorem 10.15 *Let I be an ideal in $\mathbb{C}[x_1, \ldots, x_n]$, $k \in \{1, \ldots, n-1\}$ and let I_k be the k-th elimination ideal of I. Then $V(I_k)$ is the smallest algebraic variety that contains the projection of $V(I)$ to $\mathrm{lin}\{e^{(k+1)}, \ldots, e^{(n)}\}$.*

Here, "smallest" algebraic variety is interpreted with respect to the partial order induced by set containment. That is, every algebraic variety that contains the projection of $V(I)$ is a superset of $V(I_k)$. Also, $e^{(i)}$ denotes the i-th unit vector in $\mathbb{C}^n$, and hence the projection is to the last $n - k$ coordinates.

Remark 10.16 The projection π of $V(I)$ to the coordinates $x_{k+1}, \ldots, x_n$ is not necessarily an algebraic variety. As an example, consider $f = xy - 1 \in \mathbb{C}[x, y]$. Here, $\pi(V(I))$ is the projection of the hyperbola defined by $xy = 1$, i.e., $\pi(V(I)) = \mathbb{C} \setminus \{0\}$. This set is not an algebraic variety.

Proof The projection of $V(I)$ is clearly contained in $V(I_k)$. To show that $V(I_k)$ is the smallest algebraic variety that contains the projection of $V(I)$, we study an arbitrary polynomial f in $\mathbb{C}[x_{k+1}, \ldots, x_n]$ that vanishes at all points of the projection. As a polynomial in $\mathbb{C}[x_1, \ldots, x_n]$, f vanishes at all points of $V(I)$. By the Nullstellensatz, there exists an $s \geq 1$ with $f^s \in I$. Since f^s does not depend on $x_1, \ldots, x_k$, $f^s \in I_k$. Therefore, f is contained in the radical of I_k. $\qquad\square$

We have now arrived at the main goal of the second part of this book, i.e., a method to solve systems of polynomial equations

$$f_1(x_1, \ldots, x_n) = \cdots = f_t(x_1, \ldots, x_n) = 0. \tag{10.8}$$

Here, let $f_1, \ldots, f_t \in \mathbb{C}[x_1, \ldots, x_n]$ and, as usual, set $I = \langle f_1, \ldots, f_r \rangle$. We will assume that (10.8) has only finitely many solutions, i.e., the corresponding affine variety $V(I) \subseteq \mathbb{C}^n$ is 0-dimensional. The analysis of higher dimensional varieties is beyond the scope of this book. One of the difficulties of this analysis is illustrated in Example 10.19.

Since we assumed the number of solutions to be finite, we know that for each $k \in \{1, \ldots, n-1\}$ the projection of $V(I)$ to the last $n-k$ coordinates is finite, and therefore is also an algebraic variety. By Theorem 10.15, this projection equals $V(I_k)$. For the special case $k = n-1$, this implies by Theorem 10.1 that there exists a polynomial $f \in I_{n-1}$. By Theorem 10.6, the roots of this univariate polynomial in x_n are the x_n-components of the solutions of (10.8).

Using this technique, we can employ Gröbner bases to reduce the task of solving systems of polynomial equations to the computation of roots of univariate polynomials. The computed (or approximated) roots of the univariate polynomials can then be used to combine the single components of the partial solutions to form a general solution. In addition to Example 10.2, we analyze several other examples.

Example 10.17 We use `Singular` to compute the intersection of Steiner's Roman surface (8.1) from Example 8.4 with a circle. The circle will be defined as the intersection of a sphere and a plane. The necessary steps in `Singular` are hence the following.

First, we load the library `solve.lib` which provides several functions for solving polynomial equations.

```
> LIB "solve.lib";
```

The output of this command is suppressed here (and in all of the following examples); here `Singular` would print a list of currently available libraries.

```
> ring R = 0, (x,y,z), lp;
> poly roman = x^2*y^2 + y^2*z^2 + z^2*x^2 - 2*x*y*z;
> poly sphere = x^2 + y^2 + z^2 - 1;
> poly plane = x-z;
> ideal I = roman, sphere, plane;
> ideal G = groebner(I);
> G;
```

```
G[1]=9z6-4z4
G[2]=2yz2+3z4-2z2
G[3]=y2+2z2-1
G[4]=x-z
```

The first polynomial in the output is the eliminant $9z^6 - 4z^4$ whose roots can be found using the command `laguerre_solve`.

```
> laguerre_solve(G[1]);
  [1]:
     -0.66666667
  [2]:
     0.66666667
  [3]:
     0
  [4]:
     0
  [5]:
     0
  [6]:
     0
```

The eliminant has 0 as a root of order 4 and the two order 1 roots $\pm 2/3$. By iterating the process of substituting and solving univariate polynomials, we can compute the points of the variety step by step. We illustrate an example of this procedure for the value $z = -2/3$.

```
> laguerre_solve(subst(G[2],z,-2/3));
  [1]:
     0.33333333
```

shows that for $z = -2/3$, only $y = 1/3$ is possible. Since

```
> subst(subst(G[3],z,-2/3),y,1/3);
  0
```

we see that $(1/3, -2/3)$ is actually a point of $V(I_1)$. So

```
> laguerre_solve(subst(subst(G[4],z,-2/3),y,1/3));
  [1]:
     -0.66666667
```

yields the first point $(-2/3, 1/3, -2/3)$ of the variety. In this manner, we find that the variety consists of four points, all of which are real:

$$(-2/3, 1/3, -2/3), \qquad (2/3, 1/3, 2/3), \qquad (0, -1, 0) \quad \text{and} \quad (0, 1, 0).$$

Example 10.18 In Section 8.7 we studied the intersection of two algebraic curves in the plane using `Maple`; see Fig. 8.8. The common intersection points were computed using resultants.

Since there exist only finitely many intersection points, we are able to compute, as above, an eliminant via a $\prec_{\text{lex}}$-Gröbner basis, and then iterate the partial solutions to points of the variety. In `Singular`, this approach is implemented in the library `solve.lib`.

```
> LIB "solve.lib";
> ring R = 0, (x,y), lp;
> poly f = (x^2+y^2)^2+3*x^2*y-y^3;
> poly g = y-(x^2-1);
> ideal I = f,g;
> ideal G = groebner(I);
```

The output of the following command is almost self-explanatory. For further details we refer to the `Singular` user manual.

```
> solve(G);
  [1]:
     [1]:
        -0.66555112
     [2]:
        -0.5570417
  [2]:
     [1]:
        0.66555112
     [2]:
        -0.5570417
  [3]:
     [1]:
        -0.82810567
     [2]:
        -0.314241
  [4]:
     [1]:
        0.82810567
     [2]:
        -0.314241
  [5]:
     [1]:
        (-1.32317595-i*0.90285837)
     [2]:
        (-0.064358647+i*2.389281)
  [6]:
     [1]:
        (1.32317595+i*0.90285837)
     [2]:
        (-0.064358647+i*2.389281)
  [7]:
     [1]:
        (-1.32317595+i*0.90285837)
     [2]:
        (-0.064358647-i*2.389281)
```

```
[8]:
   [1]:
      (1.32317595-i*0.90285837)
   [2]:
      (-0.064358647-i*2.389281)

// 'solve' created a ring, in which a list SOL of numbers
// (the complex solutions) is stored.
// To access the list of complex solutions, type (if the
// name R was assigned to the return value):
         setring R; SOL;
//   characteristic : 0 (complex:8 digits,
                              additional 8 digits)
//   1 parameter    : i
//   minpoly        : (i^2+1)
//   number of vars : 2
//        block    1 : ordering lp
//                   : names      x y
//        block    2 : ordering C
```

The details of the higher dimensional case are beyond the scope of this book.
However, we want to at least provide the reader with an example. It is simple to
show that a 0-dimensional, i.e., finite affine variety in $\mathbb{C}^n$ is defined by at least n
polynomials. However, the converse, even in the case where the polynomials have
no common divisor, is in general not true.

Example 10.19 We again study Steiner's Roman surface from Example 8.4. As previously mentioned, the surface has the three coordinate axes as singular loci. The
intersection of the surface with the z-axis can be modeled as the affine variety of the
ideal

$$I = \langle x^2y^2 + y^2z^2 + z^2x^2 - 2xyz, x, y \rangle \subseteq \mathbb{C}[x, y, z].$$

One can see (without using `Singular`) that $\{x, y\}$ is a $\prec_{\text{lex}}$-Gröbner basis for I.
Hence, $V(I)$ is the z-axis.

Eliminating the variables with respect to the order x, y, z is not a good idea here.
The second elimination ideal I_2 is the zero ideal, and hence we do not have an
eliminant.

If we change the order of unknowns to z, x, y, then $I_2 = \langle y \rangle$ and $I_1 = \langle x, y \rangle$
implies $x = y = 0$ for the common roots (x, y) of I_1 and I_2. Since z is then arbitrary, this example shows that the extension of the solution by Theorem 10.6 is not
necessarily unique.

10.6 Gröbner Bases and Integer Linear Programs

We conclude this chapter by discussing one of the many connections between Gröbner bases and algorithmic questions about integer point sets. This is based on the

simple fact that points with integral coordinates can be identified with monomials via their exponents. We already saw this when we discussed the Gordan–Dickson Lemma 9.20.

For $A \in \mathbb{R}^{m \times n}$, $b \in \mathbb{R}^m$ and $c \in (\mathbb{R}^n)^*$ we call

$$\min\{cx : Ax = b, x \in \mathbb{N}^n\} \tag{10.9}$$

an *integer linear program in standard form*. Note that the condition $x \in \mathbb{N}^n$ implies the non-negativity of all variables.

Remark 10.20 Similar to our comments on linear programs in Section 4.7, we remark that there are also other normal forms for integer linear programs. For our purposes, the form (10.9) seems most suitable since the inequality constraints have a rather simple structure ($x \in \mathbb{N}^n$) and the equality constraints will be particularly accessible from the viewpoint of ideals.

Conti and Traverso developed a method based on Gröbner techniques to solve arbitrary problems of the form (10.9). We discuss here the special case where $A \in \mathbb{N}^{m \times n}$, $b \in \mathbb{N}^m$ and $c \geq 0$.

The basic idea is to code the given natural numbers as exponents of polynomials. For this we define n monomials

$$f_j := w_1^{a_{1j}} \cdots w_m^{a_{mj}}, \quad \text{for } j \in \{1, \dots, n\}$$

in the polynomial ring $K[w_1, \dots, w_m]$ over an arbitrary field K. Furthermore, we define the map $\phi : K[x_1, \dots, x_n] \to K[w_1, \dots, w_m]$ via

$$\phi(x_j) := f_j$$

and its extension by $\phi(g(x_1, \dots, x_n)) = g(\phi(x_1), \dots, \phi(x_n))$. A point $\zeta \in \mathbb{N}^n$ is a feasible point of the integer program (10.9) if and only if $\phi(x^\zeta) = w^b$.

It is possible to elegantly express the property $\phi(x^\zeta) = w^b$ using the subalgebra

$$K[f_1, \dots, f_n]$$

$$:= \{p(f_1, \dots, f_n) \in K[w_1, \dots, w_m] : p \text{ polynomial with coefficients in } K\}$$

of $K[w_1, \dots, w_m]$ generated by the polynomials $f_1, \dots, f_n$. The subalgebra generated by a set of polynomials is clearly contained in the ideal generated by the same polynomials, but is usually (much) smaller: As an example, consider the subalgebra $K[1]$ generated by the constant polynomial 1. While $K[1]$ is the subalgebra of all constants in $K[w_1, \dots, w_m]$ (which is isomorphic to K itself), the ideal $\langle 1 \rangle$ is the whole polynomial ring.

Theorem 10.21 *The optimization problem* (10.9) *has a feasible solution if and only if*

$$w_1^{b_1} \cdots w_m^{b_m} \in K[f_1, \dots, f_n]. \tag{10.10}$$

The image of ϕ consists of exactly those polynomials in $K[w_1, \ldots, w_n]$ which can be expressed as polynomials in $f_1, \ldots, f_n$. To prove Theorem 10.21 we need to show that every monomial in the image of ϕ is the image of a *monomial*. For the Gröbner-based proof and the corresponding solution algorithm, let I_A be the binomial ideal

$$I_A := \langle f_1 - x_1, \ldots, f_n - x_n \rangle \subseteq K[w_1, \ldots, w_m, x_1, \ldots, x_n].$$

Here it is crucial to choose a monomial order which is suitable for our problem.

Exercise 10.22 Show that for an arbitrary monomial order $\prec$,

$$\alpha \prec_c \beta \quad :\Longleftrightarrow \quad c\alpha < c\beta \quad \text{or} \quad (c\alpha = c\beta \text{ and } \alpha \prec \beta)$$

defines a monomial order on $K[x_1, \ldots, x_n]$. Which property of monomial orders requires the non-negativity of c?

For example, if $\prec \, = \, \prec_{\text{lex}}$ and $c = \mathbf{1}$ is the all-ones vector then $\prec_c \, = \, \prec_{\text{glex}}$ is the *graded lexicographic order*, which first compares by total degree and then lexicographically to break ties.

We extend $\prec_c$ to a monomial order on the larger ring $K[w_1, \ldots, w_m, x_1, \ldots, x_n]$; here, each monomial that contains an unknown w_i has to be larger than every monomial that consists only of unknowns x_j, and furthermore, the extended monomial order has to be the lexicographic order when we restrict it to the subring $K[w_1, \ldots, w_m]$. We also denote this extension by $\prec_c$.

From now on, let $G = \{g_1, \ldots, g_t\}$ be a Gröbner basis of the ideal I_A with respect to the monomial order $\prec_c$.

Proposition 10.23 *Let $f \in K[w_1, \ldots, w_m]$ and $g = \text{rem}(f; G)$.*

(a) *We have $f \in K[f_1, \ldots, f_n]$ if and only if $g \in K[x_1, \ldots, x_n]$.*
(b) *If $f \in K[f_1, \ldots, f_n]$, then we have $f = g(f_1, \ldots, f_n)$.*

The following representation for given polynomials

$$u_1, \ldots, u_n \in K[w_1, \ldots, w_m, x_1, \ldots, x_n]$$

and $\alpha \in \mathbb{N}^n$ will be very useful in the proof:

$$
\begin{aligned}
u_1^{\alpha_1} \cdots u_n^{\alpha_n} &= \big((u_1 - x_1) + x_1\big)^{\alpha_1} \cdots \big((u_n - x_n) + x_n\big)^{\alpha_n} \\
&= v_1 \cdot (u_1 - x_1) + \cdots + v_n \cdot (u_n - x_n) + x_1^{\alpha_1} \cdots x_n^{\alpha_n} \quad (10.11)
\end{aligned}
$$

with suitable polynomials $v_1, \ldots, v_n \in K[w_1, \ldots, w_m, x_1, \ldots, x_n]$.

Proof By the definition of g, there exist $h_1, \ldots, h_t \in K[w_1, \ldots, w_m, x_1, \ldots, x_n]$ such that

$$f = h_1 g_1 + \cdots + h_t g_t + g. \quad (10.12)$$

If we assume that $g \in K[x_1, \ldots, x_n]$, then we can substitute the polynomial f_j in (10.12) for each unknown x_j. Then, since $f \in K[w_1, \ldots, w_m]$, the left hand side does not change. On the right hand side we have $g_k(f_1, \ldots, f_n) = 0$ for all k, since the generators of the Gröbner basis G are contained in the ideal $I = \langle f_1 - x_1, \ldots, f_n - x_n \rangle$. Thus, $f = g(f_1, \ldots, f_n)$, and therefore $f \in K[f_1, \ldots, f_n]$.

Conversely, let $f \in K[f_1, \ldots, f_n]$. Then, there exists a polynomial $h \in K[x_1, \ldots, x_n]$ with $f = h(f_1, \ldots, f_n)$. We need to show that $\mathrm{rem}(f; G)$ is contained in $K[x_1, \ldots, x_n]$. If we apply the trick from (10.11) to all monomials in h, multiply by the corresponding coefficients, and sum the "coefficient polynomials" v_i of the corresponding factors $(f_i - x_i)$, we obtain

$$f = h(f_1, \ldots, f_n) = p_1(f_1 - x_1) + \cdots + p_n(f_n - x_n) + h(x_1, \ldots, x_n) \quad (10.13)$$

for suitable polynomials $p_1, \ldots, p_n \in K[w_1, \ldots, w_m, x_1, \ldots, x_n]$; in particular, the difference $f - h$ is contained in the ideal I.

Now, let $G' := G \cap K[x_1, \ldots, x_n]$. Without loss of generality, we can assume that $G' = \{g_1, \ldots, g_s\}$ for an $s \leq t$. Multivariate division with remainder yields

$$h = q_1 g_1 + \cdots + q_s g_s + h' \quad (10.14)$$

for suitable polynomials $q_1, \ldots, q_s \in K[x_1, \ldots, x_n]$ and

$$h' = \mathrm{rem}(h; g_1, \ldots, g_s) \in K[x_1, \ldots, x_n].$$

Since each of the polynomials g_i is contained in the ideal I_A, by (10.13) and (10.14) there exist polynomials $q'_1, \ldots, q'_n \in K[w_1, \ldots, w_m, x_1, \ldots, x_n]$ such that

$$f = q'_1(f_1 - x_1) + \cdots + q'_n(f_n - x_n) + h'.$$

We will now show that $h' = \mathrm{rem}(f; G)$, which implies the statement. By Exercise 9.35, this is equivalent to showing that no term of h' is divisible by one of the leading terms $\mathrm{lt}(g_1), \ldots, \mathrm{lt}(g_t)$.

Now assume that $\mathrm{lt}(g_i)$ divides a term of h'. Then $\mathrm{lt}(g_i) \in K[x_1, \ldots, x_n]$, since $h' \in K[x_1, \ldots, x_n]$. Due to our special choice of the monomial order $\prec_c$, this implies $g_i \in K[x_1, \ldots, x_n]$, and thus $i \leq s$ or $g_i \in G'$, respectively. This contradicts h' being the remainder of a division by $g_1, \ldots, g_s$. $\qquad\square$

Based on this we can present the Gröbner-based method of solving the integer programming problem (10.9).

Theorem 10.24 *If $w^b = w_1^{b_1} \cdots w_m^{b_m} \in K[f_1, \ldots, f_n]$, then $\mathrm{rem}(w^b; G)$ is a monomial, i.e., $\mathrm{rem}(w^b; G) = x^\omega$ for some $\omega \in \mathbb{N}^n$, and the multi-index ω is an optimal solution of the integer linear program (10.9).*

In particular, this statement implies the criterion for the existence of a solution that was given in Theorem 10.21.

Proof Let $w^b = w_1^{b_1} \cdots w_m^{b_m} \in K[f_1, \ldots, f_n]$. By Proposition 10.23(a), we have that $\text{rem}(w^b; G)$ is contained in the subring $K[x_1, \ldots, x_n]$. By the property of binomial ideals stated in Lemma 9.30(a), $\text{rem}(w^b; G)$ is a monomial, which we denote by x^ω.

Assume there exists a $\zeta \in \mathbb{N}^n$ with $c\zeta < c\omega$. Then $\phi(x^\omega) = w^b = \phi(x^\zeta)$, and hence $\phi(x^\omega - x^\zeta) = 0$. This implies $x^\omega - x^\zeta \in I_A$, and therefore $\text{rem}(x^\omega - x^\zeta; G) = 0$. By construction of the monomial order $\prec_c$, $c\omega > c\zeta$ yields $\text{lt}(x^\omega - x^\zeta) = x^\omega$. Since $\text{rem}(x^\omega - x^\zeta; G) = 0$, x^ω must be divisible by one of the binomials $g_1, \ldots, g_t$. This contradicts the assumption that $x^\omega = \text{rem}(w^b; G)$ is the remainder of a division by $g_1, \ldots, g_t$. $\qquad\qquad\square$

Example 10.25 We examine the integer linear program $\min\{cx : Ax = b, x \in \mathbb{N}^3\}$ where

$$A = \begin{pmatrix} 2 & 1 & 1 \\ 1 & 3 & 0 \end{pmatrix}, \qquad b = \begin{pmatrix} 2 \\ 3 \end{pmatrix} \quad \text{and} \quad c = (1, 5, 2). \tag{10.15}$$

The solution set of the linear system of equations $Ax = b$ is the line

$$\begin{pmatrix} \frac{3}{5} \\ \frac{4}{5} \\ 0 \end{pmatrix} + \mathbb{R} \begin{pmatrix} -3 \\ 1 \\ 5 \end{pmatrix},$$

and the point $(0, 1, 1)$ is a feasible solution.

We can apply the Conti–Traverso method using `Singular` in the following way. First, define a ring $R = \mathbb{Q}[w_1, w_2, x_1, x_2, x_3]$ with monomial order $\prec_c$.

```
> ring R = 0, (w1,w2,x1,x2,x3), (lp(2), Wp(1,5,2));
```

The parameter `(lp(2), Wp(1,5,2))` determines the monomial order: On the subalgebra generated by the first two unknowns, i.e., $\mathbb{Q}[w_1, w_2]$, the lexicographic order is induced. On the complementary subalgebra $\mathbb{Q}[x_1, x_2, x_3]$ the monomial order from Exercise 10.22 (with the lexicographic order as secondary criterion) is induced. In general, the product order of these two monomial orders is used where the first unknowns are larger than the last unknowns in this ordering.

Next, we define the monomials f_1, f_2, f_3, and print, as a test, the leading monomial of the binomial $f_1 - x_1$:

```
> poly f1 = w1^2*w2;
> poly f2 = w1^1*w2^3;
> poly f3 = w1^1*w2^0;
> lead(f1-x1);
  w1^2*w2
```

We compute the Gröbner basis of I_A with respect to $\prec_c$, and determine the normal form of $w^b = w_1^{b_1} w_2^{b_2}$:

```
> ideal I_A = f1-x1, f2-x2, f3-x3;
> ideal G = groebner(I_A);
> G;
  G[1]=x2*x3^5-x1^3
  G[2]=w2*x1^2-x2*x3^3
  G[3]=w2*x3^2-x1
  G[4]=w2^2*x1-x2*x3
  G[5]=w2^3*x3-x2
  G[6]=w1-x3
> reduce(w1^2*w2^3, G);
  x2*x3
```

That is, the point $(0, 1, 1)$ is an optimal solution of the integer linear program given by (10.15).

Now we modify the right side to $(5, 3)$ and $(3, 1)$, and obtain optimal solutions for the modified integer linear programs.

```
> reduce(w1^5*w2^3, G);
  x2*x3^4
> reduce(w1^3*w2, G);
  x1*x3
```

Hence, the solutions are $(0, 1, 4)$ and $(1, 0, 1)$. We again modify the right hand side to $(3, 2)$ and obtain:

```
> reduce(w1^3*w2^2, G);
  w2*x1*x3
```

Since $w_2 x_1 x_3 \notin \mathbb{Q}[x_1, x_2, x_3]$, the corresponding integer linear program has no feasible solution.

Singular has specific functions for solving integer linear programs of the type (10.9).

Example 10.26 We again study Example (10.15). First, we load a library that provides specific functions for this class of problems.

```
> LIB "intprog.lib";
```

The matrix A, the right side b, and the objective function are defined as integer matrices and vectors respectively.

```
> intmat A[2][3]=2,1,1, 1,3,0;
> intvec b=2,3;
> intvec c=1,5,2;
```

Calling the function solve_IP computes the solution.

```
> print(solve_IP(A,b,c,"pct"));
  0,
  1,
  1
```

The parameter value `pct` indicates the positive version of the Conti–Traverso algorithm. For other options, we refer to the `Singular` user manual.

10.7 Exercises

Exercise 10.27 An ideal I is called a *radical ideal* if $I = \mathrm{rad}(I)$. Show that if I and J are radical ideals, then $I \cap J$ is a radical ideal.

Exercise 10.28 Sketch an alternative proof for Theorem 10.1 in which the Gröbner basis property is verified directly for the sets $G_k = G \cap K[x_{k+1}, \ldots, x_n]$.

Exercise 10.29 Sketch an alternative proof for Proposition 10.4 based on Lemma 8.9 and the extended Euclidean algorithm from Exercise 9.34.

Exercise 10.30

(a) Let K be an arbitrary *infinite* field. Show (via induction on the number of unknowns) that for every non-zero polynomial in $K[x_1, \ldots, x_n]$ there exists a point in K^n where the polynomial does not vanish.
(b) Now let K be an arbitrary *finite* field. Show that there exist for all $n \geq 1$ nonzero polynomials $f \in K[x_1, \ldots, x_n]$ such that $V_K(f) = K^n$.

10.8 Remarks

In addition to the programs `Maple`, `Sage` and `Singular` which are described in Appendix D, there exist several other software packages for the computation of Gröbner bases, including `CoCoA` [24] and `Macaulay 2` [51].

Rabinowitsch's trick to deduce the strong form of the Nullstellensatz from the weak form can be traced back to the one-page paper [87].

Further sources for algorithms and applications of Gröbner bases are the books by Cox, Little and O'Shea [28, 29], as well as Greuel and Pfister [52].

We limited our presentation of the algorithm by Conti and Traverso to a special case. Please refer to the original work [26], as well as the book by Cox, Little and O'Shea [29], for a general approach. Several further connections between Gröbner bases, convex polytopes and lattice points (e.g., the theory of toric ideals), and additional approaches to solving polynomial equations, can be found in Sturmfels' books [94, 95].

Besides its theoretical relevance, the Conti–Traverso algorithm for solving integer linear programs (ILP) is also practically important in certain special cases (in particular for series of ILPs with constant matrix A and varying right side b). However, the standard method is the *Branch-and-Bound method* or rather its refined version *Branch-and-Cut*, see Schrijver [91, §24], Korte and Vygen [72, Chapter 5], or Bertsimas and Weismantel [12].

Part III
Applications

Chapter 11
Reconstruction of Curves

There are several ways to define a curve in the plane. For example, explicitly parameterized as a continuous function $f : [0, 1] \to \mathbb{R}^2$, or (as in the case of an affine algebraic curve) implicitly as the zero set of a bivariate polynomial. For some technical applications, there are different ways of representing a curve which are more useful in those contexts, for example the representation as a Bézier curve in computer aided design (CAD). A different approach is necessary when we want to represent curves which are the result of a measurement, i.e., when they are only partially known, or the description is not exact.

We study the problem of reconstructing a curve from a given unordered set of points. Clearly, there are infinitely many curves that contain a given finite set of points. The question arises if one of these curves is privileged in any way. As an example, consider a scan of a curve drawn by hand (see Fig. 11.1). Here, the (finitely many) scanned points lie so densely that the trajectory of the curve is easy to see. A key aspect of curve reconstruction is to develop criteria for sets of points to be "sufficiently dense". For this, we introduce in the first part of this chapter the medial axis and the local feature size as intrinsic properties of a curve. After this, we define the surprisingly simple curve reconstruction method NN-Crust. The fundamental concept behind this method is the Delone subdivision of the given points.

11.1 Preliminary Considerations

We focus on the simplest case where each connected component of the curve is closed. Furthermore, we restrict ourselves to the problem of partitioning the given point set into connected components, and to order the points within each component. Then, we obtain the (re-)constructed curve by connecting the ordered points with line segments in each component. In particular, the reconstructed curve is piecewise linear.

A *closed Jordan curve* J is the image of a continuous function $f : [0, 1] \to \mathbb{R}^2$ which is homeomorphic to the circle $\mathbb{S}^1$. By the Jordan curve theorem, we know that

M. Joswig, T. Theobald, *Polyhedral and Algebraic Methods in Computational Geometry,*   181
Universitext, DOI 10.1007/978-1-4471-4817-3_11,

Fig. 11.1 A scan of a curve

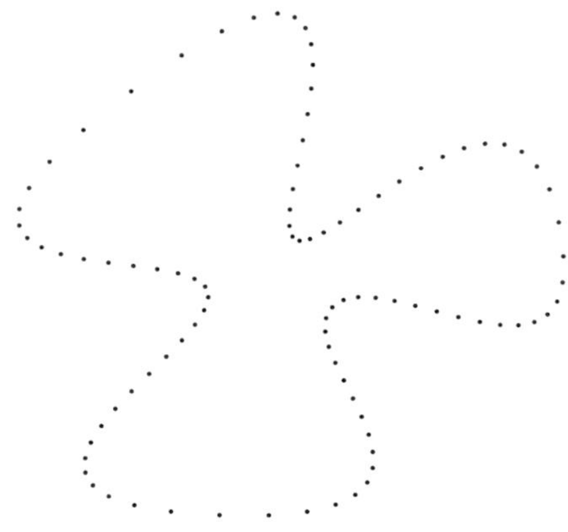

$\mathbb{R}^2 \setminus J$ has exactly two connected components, i.e., that the curve divides the plane into an inner and an outer part. A subset of J that is homeomorphic to the interval $[0, 1]$ is called a *curve arc* of J. A *sample S* on J is a finite subset of J such that $|S| \geq 3$. Two points $s^{(1)}, s^{(2)} \in S$ are called *neighbors* on J with respect to S if one of the curve arcs between $s^{(1)}$ and $s^{(2)}$ contains no other point of S. Since $|S| \geq 3$, there exists exactly one such *connecting* curve arc for each set of neighboring points $s^{(1)}$ and $s^{(2)}$.

A *polygonal reconstruction P* of a curve J is a closed polygonal chain whose vertices S form a sample on J, such that the points $s^{(1)}, s^{(2)} \in S$ are neighbors in P if and only if they are neighbors in J.

For simplicity, we stretch the common terminology by calling a union C of finitely many pairwise disjoint closed Jordan curves a *curve*. Accordingly, a sample is a union of samples of the connected components.

Remark 11.1 It seems natural to restrict our discussion to connected curves. However, the connectedness of the result of a polygonal reconstruction of a point set S depends on both the method used and the assumptions about S.

11.2 Medial Axis and Local Feature Size

In the following, let C be a curve in $\mathbb{R}^2$. We now define a continuous counterpart of the Voronoi diagram of a finite point set; compare Fig. 11.2 with Fig. 11.6 on p. 189.

Definition 11.2 The *medial axis* of C is the topological closure $M_C \subseteq \mathbb{R}^2$ of the set of points in $\mathbb{R}^2$ whose nearest point on C is not unique.

Exercise 11.3 Let J be a closed Jordan curve. Show that the interior of J contains at least one point of the medial axis M_J. Under what conditions does the outer part of J contain at least one point of M_J? When does M_J consist of exactly one point?

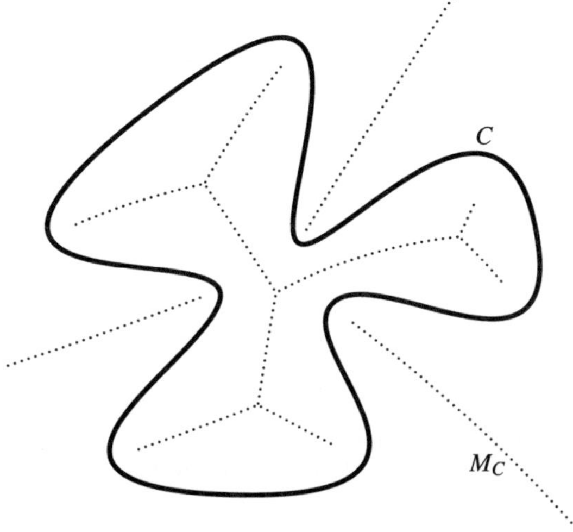

Fig. 11.2 A smooth curve with its medial axis

Throughout this chapter, we assume that the curve C is *smooth* in the sense that it is twice differentiable at every point. This implies that at each point $p \in C$ we can define the *curvature* $\kappa(p)$. Assume the connected component of p is parameterized by the function $f : [0, 1] \to \mathbb{R}^2 : t \mapsto (x(t), y(t))$. Then the curvature is defined as

$$\kappa(p) := \left| \frac{\dot{x}\ddot{y} - \ddot{x}\dot{y}}{(\dot{x}^2 + \dot{y}^2)^{3/2}} \right|.$$

Here, $\dot{x}$ denotes the derivative of x with respect to t. Just as the tangent at p is the best approximation of the curve at p by a line, the *osculating circle* to C in p is for $\kappa(p) \neq 0$ the best approximation by a circle (with common tangent). The radius of the osculating circle is $1/\kappa(p)$; see Fig. 11.3(a).

We will not need any further concepts from differential geometry for the remainder of this text. We refer the reader to the books of Kühnel [81] and Pressley [85] for more details.

Lemma 11.4 *Let $B \subseteq \mathbb{R}^2$ be a circular disk that contains at least two points of the smooth curve C. Then the intersection $B \cap C$ is homeomorphic to the interval $[0, 1]$, or B contains a point of the medial axis of C.*

Proof If $B \cap C$ is homeomorphic to $[0, 1]$, there is nothing to show. Therefore, assume that $B \cap C$ is not homeomorphic to $[0, 1]$. If one of the connected components J of C is completely contained in B, then the interior of J is also contained in B and the statement follows from Exercise 11.3.

Otherwise, $B \cap C$ is disconnected. Let z be the center of B and let p be a nearest point of z on C. We can assume that p is unique since otherwise z would be a point in M_C, and the proof would be finished. Let q be a nearest point of z on C that is not contained in the same connected component C_p of $B \cap C$ as p. Each point x

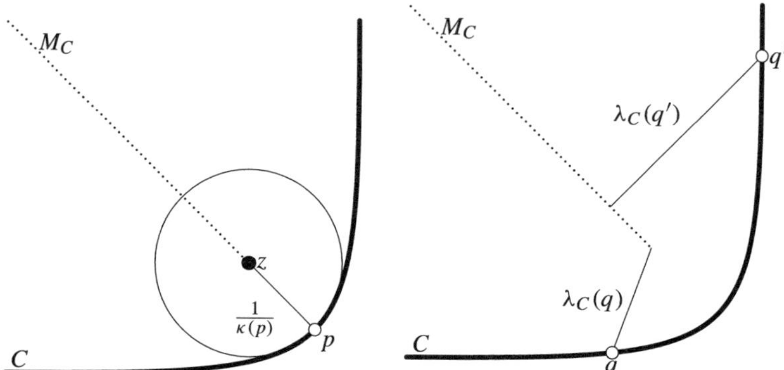

Fig. 11.3 *Left*: The center z of the osculating circle of the point p with locally maximal curvature lies on the medial axis. *Right*: The local feature size at two points q and q'

on the connecting segment of z and q is closer to q than to any point on the outside of B. Also, for each such x, the nearest point to x on C is either contained in the component C_p, or is the point q. Since z is closer to C_p than to q, the intermediate value theorem implies that there exists a point on the connecting segment of z and q that has the same distance from C_p and q. By construction, this point is contained in M_C. $\qquad\square$

We now examine the intersection of certain disks with the curve. The above lemma is crucial to this. Specifically, it states that the intersection of a disk which contains no point of the medial axis with the curve is either a curve arc, or is empty.

First, we will develop a criterion that helps to determine if the given points are sufficiently dense on a curve C to allow a polygonal reconstruction.

Definition 11.5 The *local feature size* $\lambda_C(p)$ of a curve C at the point $p \in C$ is the distance of p to the medial axis M_C.

The local feature size at a point p depends on the curvature at p, and on the other points on the curve that lie close to p.

The following statement is illustrated in Fig. 11.3(right).

Lemma 11.6 *For q and q' on C, the inequality $\lambda_C(q) \leq \lambda_C(q') + \|q - q'\|$ holds.*

Proof Let x be a point on the medial axis of C which is closest to q'. Then the triangle inequality yields that $\|q - x\| \leq \|q' - x\| + \|q - q'\| = \lambda_C(q') + \|q - q'\|$. Since, trivially, we have $\lambda_C(q) \leq \|q - x\|$, the claim follows. $\qquad\square$

We now provide a result on the intersection of curves with certain circular disks.

Lemma 11.7 *A circular disk tangent to C at the point $p \in C$ whose radius is less than or equal to $\lambda_C(p)$ contains no points of C in its interior.*

Proof Let z be the center of the largest circular disk B which lies tangent to C at p and has no points of the curve in its interior. Without loss of generality we can assume that in a neighborhood of p, B and C lie on the same side of the tangent line at p. Since B is maximal, it intersects the curve C in at least two points, or B is an osculating circle. In both cases, the center z of B is on the medial axis M_C. The local feature size at p, i.e., the distance from p to M_C, is at most as large as the distance from p to z. Therefore, the statement follows from the fact that each disk which lies tangent to C at p with a radius smaller than B is fully contained in B. $\square$

11.3 Samples and Polygonal Reconstruction

After defining samples and the local feature size, we are able to precisely state a condition under which points lie sufficiently dense on a smooth curve C, thus allowing the polygonal reconstruction.

Definition 11.8 A set $S \subseteq C$ is called an *r-sample* of C for $r \geq 0$ if there exists for each curve point p a sample point $s \in S$ such that $\|p - s\| \leq r\lambda_C(p)$.

We will see that if S is an r-sample of a curve C (for a sufficiently small r), then the edge set of a polygonal reconstruction is a subset of the Delone subdivision (Theorem 11.9). Furthermore, Lemma 11.15 will prove that the edges to the nearest neighbors of a sample point must be contained in the edge set of the polygonal reconstruction.

In the following, let S be an r-sample of the curve C. The smaller r is, the denser the points lie on the curve. As the density of the points on the curve increases, so does the precision with which one can make statements about the polygonal reconstruction of C by S. We assume that S contains at least three points of each connected component of C. In this case, there exists for each two neighboring sample points $s^{(1)}$ and $s^{(2)}$ exactly one curve arc of C with $s^{(1)}$ and $s^{(2)}$ as endpoints that contain no other points of S.

Theorem 11.9 *Let S be an r-sample of the curve C for $r < 1$, and let $s^{(1)}, s^{(2)} \in S$ be neighboring sample points on C. Then $[s^{(1)}, s^{(2)}]$ is an edge of the Delone subdivision of S and*

$$\left\| s^{(1)} - s^{(2)} \right\| \leq \frac{2r}{1 - r} \lambda_C\!\left(s^{(i)}\right), \quad 1 \leq i \leq 2. \tag{11.1}$$

If $r \leq 1/3$, in particular, the inequality $\|s^{(1)} - s^{(2)}\| \leq \lambda_C(s^{(i)})$ holds.

Proof Let p be an intersection point of the curve arc between $s^{(1)}$ and $s^{(2)}$ and the bisector of the connecting segment $[s^{(1)}, s^{(2)}]$. Also, let δ be the distance from p to the nearest sample point. Clearly, $\delta \leq \|p - s^{(1)}\| = \|p - s^{(2)}\|$. Assume that the intersection of the circular disk B around p with radius δ and the curve C is

Fig. 11.4 Sketch for the
proof of Theorem 11.9: The
construction of a circle
without interior sample points
for sufficiently dense
neighboring sample points

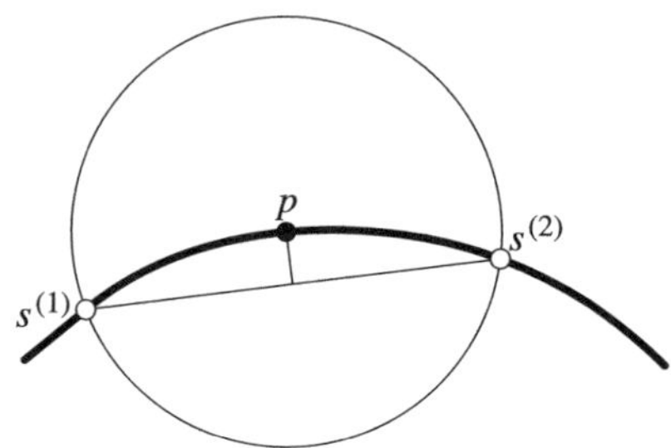

not connected. Then, by Lemma 11.4, the disk B contains a point of the medial axis. But $r < 1$ implies that there exists a sample point in the interior of B which contradicts δ being minimal. Hence, $s^{(1)}$ and $s^{(2)}$ are the sample points closest to p, i.e., $\delta = \|p - s^{(1)}\| = \|p - s^{(2)}\|$. Figure 11.4 illustrates this.

The intersection of the disk B with C is the curve arc between $s^{(1)}$ and $s^{(2)}$. In particular, B contains no sample points apart from $s^{(1)}$ and $s^{(2)}$. Theorem 7.11 states that $[s^{(1)}, s^{(2)}]$ is the edge of a Delone subdivision.

Since S is an r-sample, we have $\delta \leq r\lambda_C(p)$. The triangle inequality yields

$$\left\| s^{(1)} - s^{(2)} \right\| \leq 2\delta \leq 2r\lambda_C(p). \tag{11.2}$$

Lemma 11.6 states that $\lambda_C(p) \leq \lambda_C(s^{(i)}) + \delta \leq \lambda_C(s^{(i)}) + r\lambda_C(p)$. By (11.2) and

$$\left\| s^{(1)} - s^{(2)} \right\| \leq 2r\lambda_C(p) \leq \frac{2r}{1 - r}\lambda_C\big(s^{(i)}\big),$$

we achieve the desired inequality. $\square$

We use the assumptions and notation from Theorem 11.9. Let p be the intersection of the curve arc between $s^{(1)}$ and $s^{(2)}$ and the bisector of the connecting segment $[s^{(1)}, s^{(2)}]$.

Exercise 11.10 Show that p is the only intersection of the curve arc between $s^{(1)}$ and $s^{(2)}$ and the bisector of the connecting segment $[s^{(1)}, s^{(2)}]$.

Lemma 11.11 *The angle* $(s^{(1)}, p, s^{(2)})$ *is at least* $\pi - 2\arcsin(r/2)$.

Proof Let B be one of the two circular disks with radius $\lambda_C(p)$ that are tangent to C at p. Let z be the center of B and let $q^{(1)}, q^{(2)}$ be the intersections of the connecting segments $[s^{(i)}, z]$ with the circle ∂B. By Lemma 11.7, the curve C does not intersect the interior of B. The same is true for the point reflection of the disk at p, which is also tangent to C. We can now assume that there is no inflection point of C between $s^{(1)}$ and $s^{(2)}$. Then, the curve arc between $s^{(1)}$ and $s^{(2)}$ lies entirely on one side of the tangent line at p. Otherwise, the angle at p would increase.

The largest angle $(s^{(1)}, p, s^{(2)})$ occurs if $s^{(1)}$ and $s^{(2)}$ are on the boundary of B. We define $q^{(i)}$ as the intersection of the segment $[s^{(i)}, z]$ with ∂B for $i \in \{1, 2\}$.

Fig. 11.5 Sketch for the
proof of Lemma 11.11

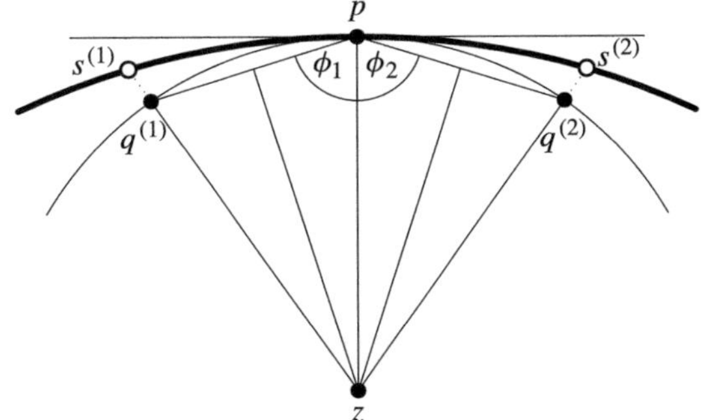

We want to provide a lower bound for the angle between $s^{(1)}$, p and $s^{(2)}$. To do
this, it suffices to give a lower bound for the sum of the angles ϕ_i between $q^{(i)}$,
p and z. Figure 11.5 provides a sketch of this. After rescaling, we can assume for
simplicity that $\lambda_C(p) = 1$.

Since $q^{(i)}$ lies between $s^{(i)}$ and z, and since $s^{(1)}$ and $s^{(2)}$ are the closest sample
points to p,

$$\left\| q^{(i)} - p \right\| \le \left\| s^{(i)} - p \right\| \le r, \quad 1 \le i \le 2. \tag{11.3}$$

Consider the two right triangles with vertices p, z and the midpoints of the segments between p and $q^{(i)}$. By (11.3), the length of the adjacent legs of ϕ_i is at
most $r/2$. Since the sine function is monotonic on the interval $[0, \pi/2]$, the inequality $\pi/2 - \phi_i \le \arcsin(r/2)$ holds. This implies

$$\pi - \phi_1 - \phi_2 \le 2\arcsin(r/2),$$

which proves the statement. $\square$

A very similar argument solves the following exercise.

Exercise 11.12 Show that for three neighboring sample points $s^{(1)}, s^{(2)}, s^{(3)}$
on C, the angle ψ between the segments $[s^{(1)}, s^{(2)}]$ and $[s^{(2)}, s^{(3)}]$ is at least
$\pi - 4\arcsin(r/2)$.

In particular, $\psi > 1.782 > \pi/2$ holds if $r \le 1/3$. [*Hint*: Consider a circular disk
which lies tangent to C at $s^{(2)}$ and has radius $\lambda_C(s^{(2)})$, and apply Lemma 11.11 to
$(s^{(1)}, s^{(2)})$ and $(s^{(2)}, s^{(3)})$.]

11.4 The Algorithm NN-Crust

Here we introduce a very simple algorithm which solves the curve reconstruction
problem for a given sample $S \subseteq \mathbb{R}^2$, if the sample is sufficiently dense on the (a pri-
ori) unknown curve C. Due to the significance of the nearest neighbors, this algorithm is called NN-Crust. The descriptions of Steps 2 and 5 in Algorithm 11.1 are
intentionally vague.

Algorithm 11.1: `NN-Crust`

    Input: finite sample $S \subseteq \mathbb{R}^2$
    Output: a subset $G(S)$ of the edges of the Delone subdivision of S
1   Compute the edge set D of the Delone subdivision of S.
2   Compute the edge set $N \subseteq D$ which connects nearest neighbors in S.
3   $G \leftarrow N$
4   **foreach** *sample point $s \in S$ that is contained in exactly one edge $e \in N$* **do**
5 | Determine the shortest edge $e' \in D$ containing s that forms an angle
      | greater than $\pi/2$ with e.
6 |_ $G \leftarrow G \cup \{e'\}$

7   **return** G

Before we analyze the conditions under which the output G of this algorithm is
a polygonal reconstruction of a curve, we analyze its complexity. Let m be the size
of the sample. We can compute the Delone subdivision for S with cost $O(m \log m)$.
Note that the size of D grows linearly in m. The following exercise shows that the
total cost of `NN-Crust` is of order $O(m \log m)$.

Exercise 11.13 Provide exact formulations of Steps 2 and 5 in the algorithm NN-
Crust which each have a cost of at most $O(m)$.

Let S be an r-sample of the curve C.

Lemma 11.14 *Let $e = [s^{(1)}, s^{(2)}]$ be the connecting segment of two non-neighboring
sample points $s^{(1)}, s^{(2)} \in S$. Then, for $i \in \{1, 2\}$ the inequality $\|s^{(1)} - s^{(2)}\| >
\lambda_C(s^{(i)})$ holds, or there exists a neighboring sample point $s' \in S$ on C for $s^{(i)}$,
such that the angle between e and $e' := [s^{(i)}, s']$ is less than or equal to $\pi/2$ and
$\|s^{(i)} - s'\| < \|s^{(1)} - s^{(2)}\|$.*

Proof Let z be the midpoint of the segment e and let B be the circular disk with
center z and diameter $\delta := \|s^{(1)} - s^{(2)}\|$. Assume that the intersection of B and C is
a curve arc. Since, by assumption, the points $s^{(1)}$ and $s^{(2)}$ are not neighbors, there
exists a third sample point $s' \in S \setminus \{s^{(1)}, s^{(2)}\}$ between $s^{(1)}$ and $s^{(2)}$ that is contained
in B. By construction, the angle between e and $e' := [s^{(1)}, s']$ is less than or equal
to $\pi/2$.

    If $B \cap C$ is disconnected, then, by Lemma 11.4, the interior of B contains a point
of the medial axis. This implies $\|s^{(1)} - s^{(2)}\| > \lambda_C(s^{(i)})$. $\square$

The following lemma determines which values of r are useful in this context.

Lemma 11.15 *Let $s^{(1)} \in S$ be an arbitrary sample point and let $s^{(2)} \in S \setminus \{s^{(1)}\}$
have minimal distance to $s^{(1)}$. For $r \leq 1/3$, the points $s^{(1)}$ and $s^{(2)}$ are neighbors on
the curve C.*

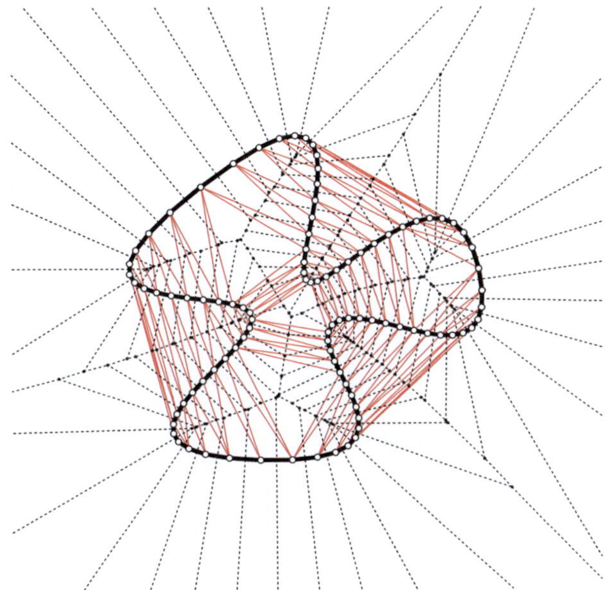

Fig. 11.6 Reconstruction of the curve in Fig. 11.2 from 96 (*white*) sample points. The *solid lines* are the edges of the Delone subdivision; the *bold edges* within the subdivision form the polygonal reconstruction. The *dashed lines* are the edges of the Voronoi diagram and the *black points* are its vertices

Proof Assume, for the sake of contradiction, that s' is a neighboring sample point of $s^{(1)}$ which is different from $s^{(2)}$. Consider the case where $\|s^{(1)} - s^{(2)}\| > \lambda_C(s^{(1)})$. Then $r \le 1/3$ and (11.1) imply

$$\left\|s^{(1)} - s'\right\| \le \frac{2r}{1-r}\,\lambda_C\!\left(s^{(1)}\right) \le \lambda_C\!\left(s^{(1)}\right).$$

From this it follows that $\|s^{(1)} - s'\| < \|s^{(1)} - s^{(2)}\|$, which contradicts our assumption that $s^{(2)}$ was a sample point with minimal distance to $s^{(1)}$.

The case $\|s^{(1)} - s^{(2)}\| \le \lambda_C(s^{(1)})$ remains to be addressed. Here, by Lemma 11.14, there exists a sample point neighboring $s^{(1)}$ which is closer to $s^{(1)}$ than $s^{(2)}$. This is again a contradiction, and thus completes the proof. $\qquad\square$

The main result of this chapter states that the algorithm NN-Crust yields the desired result for sufficiently small values of r.

Theorem 11.16 *Let S be an r-sample of the closed curve C with $r \le 1/3$. Then the algorithm* NN-Crust(S) *determines the edges of the polygonal reconstruction through S.*

Proof First we have to show that the edges which the algorithm computes connect neighboring sample points on the curve C. Secondly, we have to prove that the algorithm handles all edges.

Let $e = [s^{(1)}, s^{(2)}]$ be an edge computed by NN-Crust(S). If e was determined in Step 2, then, by Lemma 11.15, the points $s^{(1)}$ and $s^{(2)}$ are neighbors on C. Now we can assume that e was computed in Step 5. Let $x, y \in S$ be the neighboring sample points of $s^{(1)}$. Then the edge $[s^{(1)}, x]$, or the edge $[s^{(1)}, y]$ was computed in Step 2; assume $[s^{(1)}, x]$ was computed. The angle between the segments $[s^{(1)}, x]$ and $[s^{(1)}, s^{(2)}]$ is greater than $\pi/2$. By the inequality from Exercise 11.12, the angle of $[s^{(1)}, x]$ and $[s^{(1)}, y]$ is also greater than $\pi/2$. If $s^{(1)}$ and $s^{(2)}$ were not neighbors, then Lemma 11.14 would imply that $\|s^{(1)} - y\| < \|s^{(1)} - s^{(2)}\|$. This contradicts our assumption that, in Step 5, e was the shortest edge which forms an obtuse angle with $[s^{(1)}, x]$.

Conversely, assume that $s^{(1)}$ and $s^{(2)}$ are neighboring sample points on C. If $s^{(2)}$ is a sample point with minimal distance to $s^{(1)}$, then the edge $[s^{(1)}, s^{(2)}]$ will be computed in Step 2. Otherwise, Lemma 11.15 implies that the edge $[s^{(1)}, s']$ was computed in Step 2, where s' is the other neighbor of $s^{(1)}$ on C. By the inequality in Exercise 11.12, the angle between $[s^{(1)}, s']$ and $[s^{(1)}, s^{(2)}]$ is larger than $\pi/2$. Lemma 11.14 implies that $e = [s^{(1)}, s^{(2)}]$ is the shortest of all such edges. Therefore, the edge e is computed in Step 5. $\qquad\qquad\square$

11.5 Curve Reconstruction with `polymake`

As discussed in Chapter 6, we can generate objects of type `VoronoiDiagram` in `polymake`, for example, in the following way

```
polytope > $S=new Matrix([[32,99.2],[24.375,81.05],
                          [40.6,76.4],[28.925,61.975],
                          [42,52.2],[58.2,61.25],
                          [54.2,75.4],[70.35,73.9],
                          [79,85],[63.8,90.93],
                          [49.8,83.2],[51.95,102.3]]);
polytope > $m=$S->rows();
polytope > $D=new VoronoiDiagram(SITES=>ones_vector($m)|$S);
```

The 12 rows of the matrix `$S` define a point set in $\mathbb{R}^2$ that we want to reconstruct a curve from. Notice that the property `SITES` needs the point coordinates in homogenized form. The steps of the algorithm `NN-Crust` can be visualized via:

```
polytope > metapost($D->VISUAL_NN_CRUST);
```

`polymake` offers interfaces to a variety of visualization tools. The interface to METAPOST [62] creates a file (here named `D_Voronoi_Diagram.mp`) which is useful as a starting point for a high-quality planar drawing. The result is illustrated in the upper left hand corner of Fig. 11.7.

11.6 Exercises

Recall the following definition from Chapter 6: An *S-Voronoi disk* of a finite set $S \subseteq \mathbb{R}^2$ is a circular disk whose center is a vertex of the Voronoi diagram of S and whose interior contains no point of S, but its boundary contains at least one point

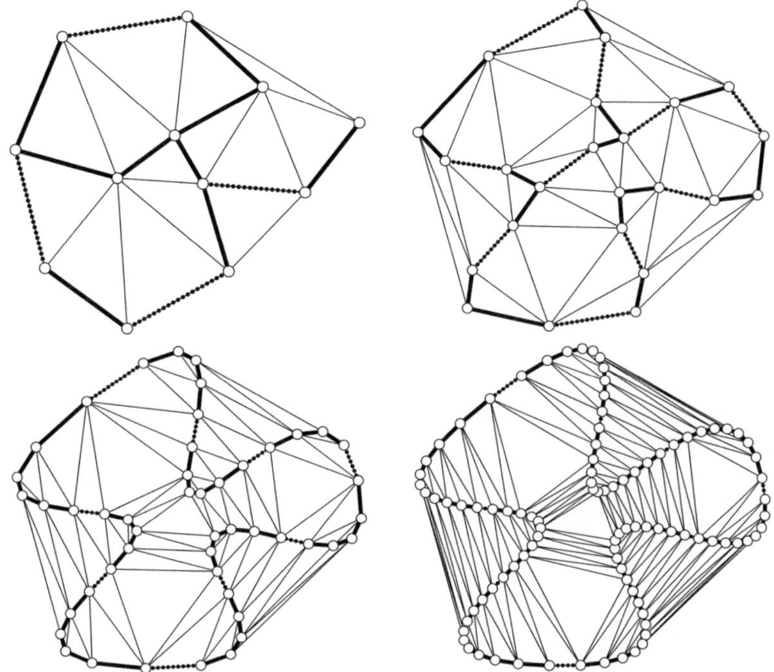

Fig. 11.7 Output of the algorithm NN-Crust for 12, 24, 48 and 96 sample points. The picture show the edges of the Delone subdivision. The edges between closest neighbors are drawn *bold* and the edges that are added in Step 5 of the algorithm are drawn as *bold dots*

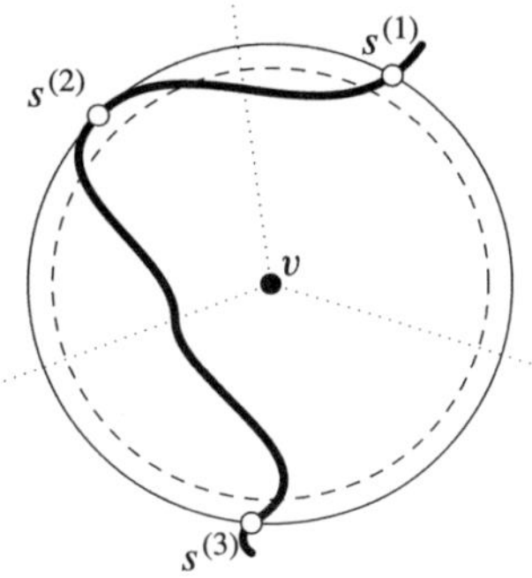

Fig. 11.8 A curve and Voronoi disk around v. The intersection of the slightly contracted disk with the curve is disconnected

of S. By Corollary 6.15, the boundary of an S-Voronoi disk contains at least three points of S.

Exercise 11.17 Let S be a finite set of points on C. Show that every S-Voronoi disk contains one point of the medial axis M_C. [*Hint*: Study Fig. 11.8.]

The next two exercises examine versions of the question of how small a disk needs to be in order to guarantee that the intersection with the curve is either empty, or a curve arc.

Exercise 11.18 Let B be a circular disk that contains a curve point $p \in C$. Show that if the diameter of B is not greater than the local feature size $\lambda_C(p)$, then B intersects the curve C in a curve arc.

Exercise 11.19 Let B be a circular disk with center z on the curve C. Show that if the radius of B is not greater than the local feature size $\lambda_C(z)$, then B intersects the curve C in a curve arc.

11.7 Remarks

In some applications we are interested in achieving smoother approximations. However, even then it is useful to start with a polygonal reconstruction. Constructions such as Bézier curves or other interpolation methods assume that the order of the given points on the curve is known.

For further background on the fundamental concepts of differential geometry described here, see the books of O'Neill [81] and Pressley [85].

The medial axis was introduced in 1967 by Blum in the context of biological forms [14].

The algorithm `NN-Crust` goes back to Dey and Kumar [34]. Our presentation of this algorithm is also based on Amenta, Bern and Eppstein [4].

Althaus and Mehlhorn [3] introduced an interesting algorithm for curve reconstruction that makes use of relations to the traveling salesman problem from combinatorial optimization.

`CGAL` provides functions for the reconstruction of curves and surfaces.

Chapter 12
Plücker Coordinates and Lines in Space

Lines, especially in $\mathbb{R}^3$, play a significant role in the modeling of geometric problems in computer graphics and machine vision. For example, a point b is visible from a point a if the line segment from a to b does not intersect another object of the scene.

Although a line is an *affine* subspace of the original space, the conditions for the intersection of lines are intrinsically *non-linear*. To illustrate this, we briefly study the problem of determining the set of lines that intersect four given lines $\ell_1, \ldots, \ell_4 \subseteq \mathbb{R}^3$. (These intersection lines are called *transversals*.) If this problem were a linear or an affine linear problem, the number of solutions would always be 0, 1, or infinite. Actually, we will see below that for lines in general position, there exist exactly two (in general complex) lines with this property.

For many problems which involve the configurations of lines, it is useful to identify the lines, in a non-linear manner, with points in a higher dimensional space. This then leads to linear intersection conditions. The so-called *Plücker coordinates* achieve this.

Before we begin to study the line configurations, we will first define Plücker coordinates for arbitrary subspaces of a projective space. We study the linear intersection conditions mentioned above in this general setting before we return to the three-dimensional case at the end of this chapter.

12.1 Plücker Coordinates

A line in projective space can be represented by two points on it. However, this representation is far from unique. It is very useful in many applications to work with a unique representation of lines. Plücker coordinates have proven to be a very elegant tool in several contexts. We define these coordinates for arbitrary subspaces of projective spaces over a field K. To simplify the notation and prevent writing unnecessary indices, in this chapter we always work with a k-dimensional linear subspace of the n-dimensional space K^n. This corresponds to the $(k-1)$-dimensional projective subspaces of the projective space $\mathbb{P}_K^{n-1}$.

M. Joswig, T. Theobald, *Polyhedral and Algebraic Methods in Computational Geometry*,   193
Universitext, DOI 10.1007/978-1-4471-4817-3_12,
© Springer-Verlag London 2013

In the following, we set $N := \binom{n}{k} - 1$. Let U be a k-dimensional subspace of K^n that is spanned by the columns of an $n \times k$-matrix L. For every subset $I \subseteq \{1, \ldots, n\}$ of cardinality k, let p_I be the $k \times k$ subdeterminant of L that is defined by the rows of I. Then the vector

$$p := (p_I)_{I \subseteq \{1, \ldots, n\}, \, |I| = k}$$

is called the vector of *Plücker coordinates* of U. Since at least one of the coordinates is non-zero, this defines a point in $\mathbb{P}^N = \mathbb{P}^N_K$.

Remark 12.1 For $k = 1$ we obtain the homogeneous coordinates of a point in $\mathbb{P}^{n-1}$. In this sense, the Plücker coordinates are a generalization of homogeneous coordinates.

Example 12.2 We examine the case $n = 4$, $k = 2$, which corresponds to lines in 3-dimensional projective space. If a line ℓ is spanned by the columns of the (4×2)-matrix

$$L = \begin{pmatrix} x_1 & y_1 \\ x_2 & y_2 \\ x_3 & y_3 \\ x_4 & y_4 \end{pmatrix},$$

then the vector p of Plücker coordinates has the six components

$$p_{i,j} = x_i y_j - x_j y_i, \quad 1 \le i < j \le 4.$$

First, we show that these coordinates are well defined. For this, let L and L' be two $k \times n$-matrices whose columns span U. From linear algebra we know that there exists a regular $k \times k$-matrix C such that $L = C \cdot L'$. Therefore, all coordinates of the Plücker vector with respect to L differ from the coordinates of the Plücker vector with respect to L' by the same factor $\det C \ne 0$. The points in $\mathbb{P}^N$ corresponding to the two Plücker vectors are hence the same.

Not every vector $p \in \mathbb{P}^N$ is the Plücker vector of a k-dimensional subspace of K^n, since the components of the vector satisfy an algebraic relation. It is necessary to understand these algebraic relations in greater detail to answer computational questions (such as: is a given vector $p \in \mathbb{P}^N$ the Plücker vector of a line?). As a central structural result, we show in the following sections that these relations can be expressed by quadratic equations. To do so, it is useful to study this definition from a more abstract viewpoint—the exterior algebra of a vector space. This enables us to express interesting properties of line configurations by Plücker coordinates in a very compact form.

12.2 Exterior Multiplication and Exterior Algebra

Let K be an arbitrary field and let V be the n-dimensional vector space K^n with the standard basis $e^{(1)}, \ldots, e^{(n)}$. For $k \in \{1, \ldots, n\}$ and indices $1 \le i_1 < \cdots < i_k \le n$ we introduce the formal symbol

$$e^{(i_1)} \wedge \cdots \wedge e^{(i_k)}$$

which we call the *exterior product of the basis vectors* $e^{(i_1)}, \ldots, e^{(i_k)}$.

We construct a new vector space $\bigwedge^k V$ from the symbols $e^{(i_1,\ldots,i_k)}$ for $i_1 < \cdots < i_k$.

Definition 12.3 For $1 \le k \le n$ we define the *k-th exterior power* $\bigwedge^k V$ as the set of formal K-linear combinations, the free K-vector space product, of the symbols

$$e^{(i_1)} \wedge \cdots \wedge e^{(i_k)}, \quad 1 \le i_1 < \cdots < i_k \le n.$$

This generating system is called the *canonical basis* of $\bigwedge^k V$. Furthermore, let $\bigwedge^0 V := K$. The (exterior) direct sum

$$\bigwedge V = \bigwedge^0 V \oplus \bigwedge^1 V \oplus \cdots \oplus \bigwedge^n V$$

is called the *exterior algebra over V*.

Since there exist $\binom{n}{k}$ index sequences $1 \le i_1 < \cdots < i_k \le n$,

$$\dim_K \bigwedge^k V = \binom{n}{k} \quad \text{and} \quad \dim_K \bigwedge V = 2^n.$$

Usually, $\bigwedge^1 V$ is identified with the vector space V itself. Furthermore, we define all vector spaces $\bigwedge^k V$ with $k > n$ as the zero space.

The term "algebra" suggests that there exists a multiplication on $\bigwedge V$. Before we define the multiplication, we analyze an example.

Example 12.4 Let $V = K^4$. The second exterior power $\bigwedge^2 V$ has the canonical basis

$$e^{(1)} \wedge e^{(2)}, \qquad e^{(1)} \wedge e^{(3)}, \qquad e^{(1)} \wedge e^{(4)},$$
$$e^{(2)} \wedge e^{(3)}, \qquad e^{(2)} \wedge e^{(4)}, \qquad e^{(3)} \wedge e^{(4)},$$

$\bigwedge^3 V$ has the canonical basis

$$e^{(1)} \wedge e^{(2)} \wedge e^{(3)}, \qquad e^{(1)} \wedge e^{(2)} \wedge e^{(4)}, \qquad e^{(1)} \wedge e^{(3)} \wedge e^{(4)}, \qquad e^{(2)} \wedge e^{(3)} \wedge e^{(4)},$$

and $\bigwedge^4 V$ has the canonical basis $e^{(1)} \wedge e^{(2)} \wedge e^{(3)} \wedge e^{(4)}$. Therefore, the K-vector space dimension of the exterior algebra $\bigwedge V$ is $1 + 4 + 6 + 4 + 1 = 16 = 2^4$.

Now we define the following *exterior multiplication*, denoted by $\wedge$, on a pair of basis vectors of V via

$$e^{(i)} \wedge e^{(i)} := 0 \quad \text{and} \quad e^{(j)} \wedge e^{(i)} := -\left(e^{(i)} \wedge e^{(j)}\right) \tag{12.1}$$

for $i < j$. This map has a unique associative extension to the set of all pairs of canonical basis vectors of $\bigwedge V$ such that

$$e^{(i_1)} \wedge \cdots \wedge e^{(i_k)} = \mathrm{sgn}(\sigma) \cdot \left(e^{(\sigma(i_1))} \wedge \cdots \wedge e^{(\sigma(i_k))} \right),$$

where $\sigma \in \mathrm{Sym}(\{i_1, \ldots, i_k\})$ is a permutation.

Example 12.5 If $n \geq 3$, then $e^{(1)} \wedge e^{(2)} \wedge e^{(3)} = -(e^{(2)} \wedge e^{(1)} \wedge e^{(3)})$.

Exercise 12.6 Show that the exterior multiplication defined on pairs of canonical basis vectors has a unique extension to a K-bilinear map

$$\wedge : \bigwedge V \times \bigwedge V \to \bigwedge V,$$

the *exterior multiplication* on V.

For a systematic approach, we need to first prove some general properties of the exterior multiplication:

Exercise 12.7 Show:

(a) The exterior multiplication is associative.
(b) The exterior multiplication is skew-symmetric, i.e., $x \wedge y = -y \wedge x$ for all $x, y \in V$.
(c) For $x^{(1)}, \ldots, x^{(k)} \in V$ we have $x^{(1)} \wedge \cdots \wedge x^{(k)} = 0$ if and only if $x^{(1)}, \ldots, x^{(k)}$ are linearly dependent over K.

The relevance of the exterior algebra for Plücker coordinates, and thus for configurations of subspaces, is based on the following lemma.

Lemma 12.8 *Let U be the k-dimensional subspace of V spanned by $u^{(1)}, \ldots, u^{(k)}$. Then the vector of coefficients $p_{i_1,\ldots,i_k}$, where $i_1 < \cdots < i_k$, in the basis representation*

$$u^{(1)} \wedge \cdots \wedge u^{(k)} = \sum_{i_1 < \cdots < i_k} p_{i_1,\ldots,i_k} \cdot e^{(i_1)} \wedge \cdots \wedge e^{(i_k)} \tag{12.2}$$

of the exterior product $u^{(1)} \wedge \cdots \wedge u^{(k)}$ equals the Plücker coordinates of U in homogeneous coordinates, i.e., both vectors denote the same point in $\mathbb{P}^N$.

Proof By repeated use of linearity, we have

$$u^{(1)} \wedge \cdots \wedge u^{(k)} = \sum_{i_1,\ldots,i_k \in \{1,\ldots,n\}} u_{i_1}^{(1)} \cdots u_{i_k}^{(k)} \cdot e^{(i_1)} \wedge \cdots \wedge e^{(i_k)}. \tag{12.3}$$

Then the skew-symmetry from (12.1) gives the coefficient

$$p_{i_1,\ldots,i_k} = \sum_{\sigma \in \mathrm{Sym}(\{i_1,\ldots,i_k\})} \mathrm{sgn}(\sigma) u_{\sigma(i_1)}^{(i_1)} \cdots u_{\sigma(i_k)}^{(i_k)}$$

in the basis representation (12.3), where $\mathrm{sgn}(\sigma)$ denotes the sign of the permutation σ. Using the Leibniz expansion of the determinant, we recognize this expression for $p_{i_1,\dots,i_k}$ as the Plücker coordinate with index $\{i_1, \dots, i_k\}$. $\qquad\square$

We call the representation of Plücker coordinates in Lemma 12.8 via the exterior product the *exterior Plücker representation*.

Now we will describe when a given element $\omega \in \bigwedge^k V$ is the Plücker vector of a k-dimensional subspace, i.e., when $v^{(1)}, \dots, v^{(k)} \in V$ exist such that $\omega = v^{(1)} \wedge \cdots \wedge v^{(k)}$. To do this, fix $\omega \in \bigwedge^k V$ and examine the linear map

$$\wedge_\omega : V \to \bigwedge^{k+1} V,$$
$$v \mapsto v \wedge \omega.$$

By choosing the canonical basis for V and $\bigwedge^{k+1} V$ in lexicographic order, we obtain the corresponding representation matrix

$$M_\omega \in K^{\binom{n}{k+1} \times n}.$$

Lemma 12.9 *Let* $\omega \in \bigwedge^k V \setminus \{0\}$*, then the following properties are equivalent:*

(a) *There exist* $v^{(1)}, \dots, v^{(k)} \in V$ *with* $\omega = v^{(1)} \wedge \cdots \wedge v^{(k)}$.
(b) $\dim \ker \wedge_\omega = k$.
(c) $\mathrm{rank}\, M_\omega = n - k$.

Proof We show that (a) is equivalent to (b).

For any vector v and linearly independent $v^{(1)}, \dots, v^{(k)} \in V$, by Exercise 12.7(c)

$$v \wedge v^{(1)} \wedge \cdots \wedge v^{(k)} = 0 \quad \Longleftrightarrow \quad v \in \mathrm{lin}\big\{v^{(1)}, \dots, v^{(k)}\big\}. \qquad (12.4)$$

If $0 \neq \omega = v^{(1)} \wedge \cdots \wedge v^{(k)}$, then the vectors $v^{(1)}, \dots, v^{(k)}$ are linearly independent, and $\ker \wedge_\omega = \mathrm{lin}\{v^{(1)}, \dots, v^{(k)}\}$, hence $\dim \ker \wedge_\omega = k$.

To prove the converse, let $v^{(1)}, \dots, v^{(n)}$ be a basis of V such that the first k vectors $v^{(1)}, \dots, v^{(k)}$ are a basis of the kernel of $\wedge_\omega$. The set of vectors $v^{(I)} = v^{(i_1)} \wedge \cdots \wedge v^{(i_k)}$ with $I = \{i_1, \dots, i_k\}$ and $1 \leq i_1 < \cdots < i_k \leq n$ is a basis for $\bigwedge^k V$. Therefore, there exists a unique representation of ω as a linear combination of the basis vectors

$$\omega = \sum_I \omega_I v^{(I)}$$

with coefficients $\omega_I \in K$. For every $i \in \{1, \dots, k\}$, by construction $v^{(i)} \wedge \omega = 0$, and therefore, by the definition of the exterior product, all ω_I with $i \notin I$ vanish. As a consequence, only the coefficient $\omega_{\{1,\dots,k\}}$ can be non-zero.

The equivalence of (b) and (c) is clear. $\qquad\square$

The next exercise is an alternative version of the previous lemma.

Exercise 12.10 Let $\omega \in \bigwedge^k V \setminus \{0\}$. Show that:

$$\dim \ker \wedge_\omega = k \quad \Longleftrightarrow \quad \dim \ker \wedge_\omega \geq k.$$

We can now justify the term "coordinates" by showing that any two different k-dimensional subspaces have different Plücker vectors.

The set of k-dimensional subspaces of K^n is denoted by $G_{k,n}\,K$ and is called the k-th *Grassmannian* of K^n. We have that $G_{1,n}\,K$ and $G_{2,n}\,K$ are the set of points and lines of the projective space $\mathbb{P}_K^{n-1}$.

Lemma 12.11 *The map from the Grassmannian $G_{k,n}\,K$ to $\mathbb{P}_K^N$ that maps a k-dimensional subspace to its Plücker coordinates is injective.*

Proof Let $\omega = v^{(1)} \wedge \cdots \wedge v^{(k)}$ and $\omega' = w^{(1)} \wedge \cdots \wedge w^{(k)}$ be exterior Plücker representations (hence, in particular, non-zero). We have to show that $\mathrm{lin}\{v^{(1)}, \ldots, v^{(k)}\} = \mathrm{lin}\{w^{(1)}, \ldots, w^{(k)}\}$ if and only if ω' is a non-zero multiple of ω.

Assume, first, that $\mathrm{lin}\{v^{(1)}, \ldots, v^{(k)}\} = \mathrm{lin}\{w^{(1)}, \ldots, w^{(k)}\}$, which implies that every vector $w^{(i)}$ has a representation as $w^{(i)} = \sum_{j=1}^k \lambda_{ij} v^{(j)}$. Therefore, we have

$$\omega' = \sum_{j_1, \ldots, j_k} \lambda_{1,j_1} \cdots \lambda_{k,j_k} \cdot v^{(1)} \wedge \cdots \wedge v^{(k)}.$$

Only those terms for which $\{j_1, \ldots, j_k\}$ is a permutation of $\{1, \ldots, k\}$ can be non-zero, and thus we obtain, as in Lemma 12.8,

$$\omega' = \det\left(v^{(1)}, \ldots, v^{(k)}\right) \cdot v^{(1)} \wedge \cdots \wedge v^{(k)},$$

so that ω' is a multiple of ω.

The converse follows from Lemma 12.8 and from the fact that the Plücker coordinates are well defined. $\qquad\square$

Example 12.12 We again study the case $n = 4$, $k = 2$ for an illustration. Each vector $\omega \in \bigwedge^2 V$ has a representation of the form

$$\omega = \sum_{1 \leq i < j \leq 4} p_{ij} \cdot e^{(i)} \wedge e^{(j)}$$

with Plücker coordinates p_{ij}. As we saw before, the columns of the representation matrix M_ω of $\wedge_\omega$ consist of the coordinate vectors of the images of the canonical basis vectors. For the order $e^{(1)}, \ldots, e^{(4)}$ of the canonical basis vectors of V, and the order $e^{(1)} \wedge e^{(2)} \wedge e^{(3)}, e^{(1)} \wedge e^{(2)} \wedge e^{(4)}, e^{(1)} \wedge e^{(3)} \wedge e^{(4)}, e^{(2)} \wedge e^{(3)} \wedge e^{(4)}$ of the canonical basis vectors of $\bigwedge^3 V$, we obtain the representation matrix M_ω of $\wedge_\omega$ as

$$M_\omega = \begin{pmatrix} p_{23} & -p_{13} & p_{12} & 0 \\ p_{24} & -p_{14} & 0 & p_{12} \\ p_{34} & 0 & -p_{14} & p_{13} \\ 0 & p_{34} & -p_{24} & p_{23} \end{pmatrix}.$$

By Lemma 12.9, the vector ω defines the Plücker vector of a line in $\mathbb{P}^3$ if and only if this matrix has rank 2.

12.3 Duality

Duality is also important in the context of Plücker coordinates. When we defined the Plücker coordinates, we described subspaces U of $V = K^n$ as the span of k linearly independent vectors. If we describe U as an intersection of $n - k$ hyperplanes we obtain the dual Plücker coordinates which will be defined in the following.

Let U be a k-dimensional subspace in K^n which is given as an intersection of $n - k$ hyperplanes,

$$\sum_{i=1}^{n} u_i^{(1)} x_i = 0, \qquad \ldots, \qquad \sum_{i=1}^{n} u_i^{(n-k)} x_i = 0,$$

whose coefficient vectors $u^{(1)}, \ldots, u^{(n-k)}$ define the rows of an $(n - k) \times n$-matrix M. For every subset $I \subseteq \{1, \ldots, n\}$ of cardinality $n - k$, let q_I be the $(n - k) \times (n - k)$ subdeterminant of M that is defined by the columns of I. Then the vector in $\mathbb{P}^N$ defined by

$$q := (q_I)_{I \subseteq \{1,\ldots,n\}, \, |I|=n-k}$$

is called the vector of *dual Plücker coordinates* of U. Analogous to the representation of primal Plücker coordinates, it is possible to represent the dual Plücker coordinates as an exterior product. The dual basis $(e^{(1)})^*, \ldots, (e^{(n)})^*$ of the standard basis consists of the linear forms $x \mapsto x_1, \ldots, x \mapsto x_n$. As previously done, we can define the exterior algebra $\bigwedge V^*$ on the dual vector space V^*.

Remark 12.13 For $k = n - 1$ the dual Plücker coordinates are the homogeneous coordinates of hyperplanes in $\mathbb{P}^{n-1}$. This is the dual version of Remark 12.1.

The dual Plücker coordinates are closely related to the primal Plücker coordinates. To study this connection, it is convenient to use the compact notation

$$e^{(I)} := e^{(i_1)} \wedge \cdots \wedge e^{(i_k)}$$

for any index set $I = \{i_1, \ldots, i_k\}$ with $1 \le i_1 < \cdots < i_k \le n$. We define the operator $*$ as a linear map $\bigwedge^k V \to \bigwedge^{n-k} V^*$ by defining the images of the basis elements $e^{(I)}$ of $\bigwedge^k V$ (and by linear extension). Let $I = \{i_1, \ldots, i_k\}$ with $1 \le i_1 < \cdots < i_k \le n$, and $J = \{j_1, \ldots, j_{n-k}\} := \{1, \ldots, n\} \setminus I$ with increasing indices $j_1 < \cdots < j_{n-k}$ be the complement of I. Then we define

$$*\big(e^{(I)}\big) := \mathrm{sgn}(i_1, \ldots, i_k, j_1, \ldots, j_{n-k}) \cdot \big(e^{(J)}\big)^*$$

$$= \mathrm{sgn}(i_1, \ldots, i_k, j_1, \ldots, j_{n-k}) \cdot \big(e^{(j_1)}\big)^* \wedge \cdots \wedge \big(e^{(j_{n-k})}\big)^*.$$

Note that here and in the following, we write the permutation $(1 \mapsto i_1, \ldots, n \mapsto i_n)$ as the vector of the images $(i_1, \ldots, i_n)$.

Example 12.14 For $n = 4$, $k = 2$ the $*$ operator yields $*(1) = e^{(1234)}$, $*(e^{(1234)}) = 1$ and

$$
\begin{aligned}
*e^{(1)} &= \left(e^{(234)}\right)^*, & *e^{(12)} &= \left(e^{(34)}\right)^*, & *e^{(123)} &= \left(e^{(4)}\right)^*, \\
*e^{(2)} &= -\left(e^{(134)}\right)^*, & *e^{(13)} &= -\left(e^{(24)}\right)^*, & *e^{(124)} &= -\left(e^{(3)}\right)^*, \\
*e^{(3)} &= \left(e^{(124)}\right)^*, & *e^{(14)} &= \left(e^{(23)}\right)^*, & *e^{(134)} &= \left(e^{(2)}\right)^*, \\
*e^{(4)} &= -\left(e^{(123)}\right)^*, & *e^{(23)} &= \left(e^{(14)}\right)^*, & *e^{(234)} &= -\left(e^{(1)}\right)^*, \\
& & *e^{(24)} &= -\left(e^{(13)}\right)^*, \\
& & *e^{(34)} &= \left(e^{(12)}\right)^*.
\end{aligned}
$$

We show that the dual exterior Plücker representation of the subspace U equals (up to a multiplicative constant) the primal exterior Plücker representation after applying the $*$ operator. For this, we need the following determinant identity of Jacobi.

Lemma 12.15 *Let $A \in K^{n \times n}$ be invertible and of the form*

$$
A = \begin{pmatrix} A_{11} & A_{12} \\ A_{21} & A_{22} \end{pmatrix}, \qquad B := A^{-1} = \begin{pmatrix} B_{11} & B_{12} \\ B_{21} & B_{22} \end{pmatrix}
$$

with $k \times k$-matrices A_{11}, B_{11}. Then

$$
\det B_{22} \cdot \det A = \det A_{11}.
$$

Proof Since $A \cdot A^{-1} = \mathrm{Id}$ we have

$$
\begin{pmatrix} A_{11} & A_{12} \\ A_{21} & A_{22} \end{pmatrix} \cdot \begin{pmatrix} \mathrm{Id} & B_{12} \\ 0 & B_{22} \end{pmatrix} = \begin{pmatrix} A_{11} & 0 \\ A_{21} & \mathrm{Id} \end{pmatrix}.
$$

By computing the determinant on both sides, we immediately obtain the result. $\square$

Theorem 12.16 *Let p and q be the vectors of the primal and dual Plücker coordinates of a k-dimensional subspace U of V. If we interpret p and q as vectors in $\mathbb{R}^{N+1}$, then there exists a constant $c \neq 0$ such that for all permutations $(i_1, \ldots, i_n) \in \mathrm{Sym}(\{1, \ldots, n\})$*

$$
p_{i_1, \ldots, i_k} = c \cdot \mathrm{sgn}(i_1, \ldots, i_n) \cdot q_{i_{k+1}, \ldots, i_n}. \tag{12.5}
$$

Proof We begin with the special case of the k-dimensional subspace defined by $x_{k+1} = \cdots = x_n = 0$. This is spanned by the unit vectors $e^{(1)}, \ldots, e^{(k)}$. The coordinate $p_{i_1, \ldots, i_k}$ is non-zero if and only if $\{i_1, \ldots, i_k\} = \{1, \ldots, k\}$, and in this case $p_{i_1, \ldots, i_k}$ is 1 if and only if the permutation $(i_1, \ldots, i_k)$ has a positive sign. The same holds for $q_{i_{k+1},\ldots,i_n}$, and hence the statement follows from $\mathrm{sgn}(i_1,\ldots,i_k) \cdot \mathrm{sgn}(i_{k+1},\ldots,i_n) = \mathrm{sgn}(i_1,\ldots,i_n)$.

For the general case, we assume that the k-dimensional subspace U can be obtained from the special subspace by a linear map with representation matrix M. By Jacobi's determinant identity, Lemma 12.15, the proportionality between the primal and dual Plücker coordinates remains. $\qquad\square$

Corollary 12.17 *Let $\omega \in \bigwedge^k V$ be the exterior Plücker representation of a k-dimensional subspace U of V, then $*(\omega) \in \bigwedge^{n-k} V^*$ is an exterior representation of the dual Plücker coordinates of U.*

Corollary 12.17 and $*(*(\omega)) = (-1)^{k(n-k)}\omega$ yield:

Corollary 12.18 *An element $\omega \in \bigwedge^k V$ is an exterior Plücker representation of a k-dimensional subspace of V if and only if $*(\omega)$ is a dual exterior Plücker representation of a k-dimensional subspace of V.*

Analogous to $\wedge_\omega$, we define the map

$$\wedge_{*(\omega)} : V^* \to \bigwedge^{n-k+1} V^*,$$

$$\phi \mapsto \phi \wedge *(\omega).$$

The representation matrix of this map (with respect to the lexicographically ordered canonical basis) is denoted by $M_\omega^* \in K^{\binom{n}{n-k+1}\times n}$.

Example 12.19 In the case $n = 4$, $k = 2$, and for $\omega = \sum_{1\leq i<j\leq 4} p_{ij}(e_i \wedge e_j)$, we have

$$*(\omega) = p_{12}\bigl(e^{(34)}\bigr)^* - p_{13}\bigl(e^{(24)}\bigr)^* + p_{14}\bigl(e^{(23)}\bigr)^*$$
$$+ p_{23}\bigl(e^{(14)}\bigr)^* - p_{24}\bigl(e^{(13)}\bigr)^* + p_{34}\bigl(e^{(12)}\bigr)^*.$$

This implies in particular

$$\bigl(e^{(1)}\bigr)^* \wedge *(\omega) = p_{14}\bigl(e^{(123)}\bigr)^* - p_{13}\bigl(e^{(124)}\bigr)^* + p_{12}\bigl(e^{(134)}\bigr)^*,$$

which gives the first column of the representation matrix

$$M_\omega^* = \begin{pmatrix} p_{14} & p_{24} & p_{34} & 0 \\ -p_{13} & -p_{23} & 0 & p_{34} \\ p_{12} & 0 & -p_{23} & -p_{24} \\ 0 & p_{12} & p_{13} & p_{14} \end{pmatrix}.$$

Theorem 12.20 *An element $\omega \in \bigwedge^k V \setminus \{0\}$ is an exterior Plücker representation of a k-dimensional subspace of V if and only if*

$$M_\omega \cdot \bigl(M_\omega^*\bigr)^T = 0. \tag{12.6}$$

Proof By Corollary 12.18, the vector ω is an exterior Plücker representation of a k-dimensional subspace if and only if $*(\omega)$ is a dual exterior Plücker representation of a k-dimensional subspace. Therefore, in this case there exists a basis $v^{(1)}, \ldots, v^{(n)}$ of V such that

$$\omega = v^{(1)} \wedge \cdots \wedge v^{(k)} \quad \text{and} \quad *(\omega) = \left(v^{(k+1)}\right)^* \wedge \cdots \wedge \left(v^{(n)}\right)^*.$$

For each $v \in V$, the linear form $u \wedge *(\omega)$ vanishes on $v \wedge \omega$, which implies the stated property.

Now we study the converse. By Lemma 12.9 and Exercise 12.10, the property $\dim \ker \wedge_\omega \le k$, and analogously $\dim \ker \wedge_{*(\omega)} \le n - k$, holds for every $\omega \in \bigwedge^k V \setminus \{0\}$. Therefore, if (12.6) is satisfied we must have equality in both cases. By Lemma 12.9, the vector ω is an exterior Plücker representation. $\qquad \square$

As a corollary we now obtain the desired characterization of those points $p \in \mathbb{P}^N$ which are Plücker coordinates of a k-dimensional subspace of $V = K^n$.

Theorem 12.21 *The Plücker coordinates* $(p_I)_{I \subseteq \{1,\ldots,n\}, |I|=k}$ *of the k-dimensional subspaces of V correspond to those points of $\mathbb{P}^N$ which satisfy the condition*

$$\sum_{l=1}^{k+1} (-1)^l p_{i_1,\ldots,\hat{i}_l,\ldots,i_{k+1}} \, p_{j_1,\ldots,j_{k-1},i_l} = 0 \tag{12.7}$$

for all $i_1, \ldots, i_{k+1}, j_1, \ldots, j_{k-1} \in \{0, \ldots, n\}$, where $\hat{i}_l$ denotes that the index i_l is omitted.

Proof The representation matrix $M_\omega \in K^{\binom{n}{k+1} \times n}$ satisfies

$$(M_\omega)_{Ij} = \begin{cases} 0 & \text{if } j \notin I, \\ \epsilon p_{I \setminus \{j\}} & \text{if } j \in I, \end{cases}$$

where $I \setminus \{j\} = \{i_1, \ldots, i_k\}$ with $i_1 \le \cdots \le i_k$, and

$$\epsilon = \operatorname{sgn}(j, i_1, \ldots, i_k).$$

Analogously, $M_\omega^* \in K^{\binom{n}{n-k+1} \times n}$ satisfies

$$\left(M_\omega^*\right)_{I'j} = \begin{cases} 0 & \text{if } j \notin I', \\ \epsilon' p_J & \text{if } j \in I' \end{cases}$$

with $I' = \{i_1', \ldots, i_{n-k+1}'\}$, $i_1' \le \cdots \le i_{n-k+1}'$, $J = \{1, \ldots, n\} \setminus I' = \{j_1, \ldots, j_{k-1}\}$, $j_1 \le \cdots \le j_{k-1}$ and $\epsilon' = \operatorname{sgn}(i_1', \ldots, i_{n-k+1}', j_1, \ldots, j_{k-1})$. This yields (12.7). $\qquad \square$

For the special case $k = 2$, i.e., the case of lines in projective space, we obtain the following corollary.

Corollary 12.22 *The Plücker coordinates of a line ℓ in $\mathbb{P}^{n-1}$ satisfy the following conditions.*

$$p_{ij}p_{rs} - p_{ir}p_{js} + p_{is}p_{jr} = 0, \quad \text{for } 1 \le i < j < r < s \le n.$$

For $n = 4$ these conditions reduce to a single quadratic equation

$$p_{12}p_{34} - p_{13}p_{24} + p_{14}p_{23} = 0. \tag{12.8}$$

The quadric in $\mathbb{P}^5$ defined by (12.8) is called the *Klein quadric*. We summarize the statements of this section in the following way:

Corollary 12.23 *The map from $\mathrm{G}_{k,n}\,K$ to the variety in $\mathbb{P}^N$ defined by (12.7) which maps a subspace to its Plücker vector is bijective.*

12.4 Computations with Plücker Coordinates

The reason for introducing primal and dual Plücker coordinates is that they allow for the computation of the intersections of subspaces in a very comfortable way. Proposition 2.5 showed how the incidence relation of points and hyperplanes in projective space can be expressed using the inner product of homogeneous coordinate vectors. We generalize this now.

The inner product of two points in the projective space $\mathbb{P}^{n-1}$ (as the inner product of the representatives in K^n) is defined only up to a non-zero multiplicative constant; but analogously to Proposition 2.5, determining if the inner product vanishes is independent of the choice of representatives.

Theorem 12.24 *A $(k-1)$-dimensional projective subspace U of $\mathbb{P}^{n-1}$ intersects an $(n-k-1)$-dimensional projective subspace W of $\mathbb{P}^{n-1}$ if and only if the inner product of the Plücker coordinates p of U and the dual Plücker coordinates q of W vanishes, i.e., if*

$$\sum_{I\subseteq\{1,\ldots,n\},|I|=k} p_I q_I = 0. \tag{12.9}$$

Proof We identify projective subspaces of $\mathbb{P}^{n-1}$ with linear subspaces of $V = K^n$.

Let $u^{(1)},\ldots,u^{(k)}$ be a basis of the linear subspace U of V, and let $w^{(1)},\ldots,w^{(k)}$ be the coefficient vectors of the equations defining the linear subspace W. A point $\sum_{i=1}^{k}\lambda_i u^{(i)} \in U$ with coefficients $\lambda_1,\ldots,\lambda_k$ is contained in W if and only if

$$\sum_{i=1}^{k}\sum_{l=1}^{k}\lambda_l u_l^{(i)} w_l^{(j)} = 0, \quad \text{for all } j \in \{1,\ldots,k\}.$$

This system of equations has exactly one non-trivial solution in $\lambda_1, \ldots, \lambda_k$ if

$$\det\left(\sum_{l=1}^{n} u_l^{(i)} w_l^{(j)}\right)_{\substack{1 \le i \le k \\ 1 \le j \le k}} = 0. \tag{12.10}$$

This determinant can also be interpreted as the determinant of the product of the matrices $(u_l^{(i)})_{l,i}$ and $(w_l^{(j)})_{j,l}$.

The Cauchy–Binet formula says that for two arbitrary matrices $A \in K^{n \times k}$, $B \in K^{k \times n}$ we have

$$\det AB = \sum_{I \subseteq \{1,\ldots,n\}, |I|=k} \det A_I \det B_I, \tag{12.11}$$

where A_I and B_I are the submatrices of A and B in which only those columns of A and rows of B are used whose indices are in I. (A very elegant proof of this statement can be found in THE BOOK [2].) Using the Cauchy–Binet formula (12.11), we can write (12.10) as

$$\sum_{I} p_I q_I,$$

which implies the statement. $\square$

12.5 Lines in $\mathbb{R}^3$

Lines in three-dimensional space occur, for example, in *ray shooting* problems, in computer graphics. In the simplest situation we are given a line and a polytope in $\mathbb{R}^3$ (in computer graphics often a polygon), and we want to test if the line and the polytope intersect. Or, for a directed line ℓ and a finite set of disjoint polytopes, we have to determine the order in which ℓ intersects the polytopes. There are numerous applications of this kind.

From the viewpoint of non-linear geometry, it is especially interesting if the line $\ell \subseteq \mathbb{R}^3$ is tangent to a polytope $P \subseteq \mathbb{R}^3$. In this case, there exists a point p on an edge e of P that is contained in the line ℓ. If ℓ' denotes the line containing the edge e, we have the same situation as in Theorem 12.24. In 3-dimensional projective space with homogeneous coordinates $x_1, \ldots, x_4$ the intersection condition can be stated as follows:

Corollary 12.25 *A line ℓ intersects a line ℓ' in $\mathbb{P}^3$ if their Plücker coordinates p and p' satisfy*

$$p_{12}p'_{34} - p_{13}p'_{24} + p_{14}p'_{23} + p_{23}p'_{14} - p_{24}p'_{13} + p_{34}p'_{12} = 0. \tag{12.12}$$

With the elimination techniques developed in Section 10.2, we can simply let Singular compute the Plücker relation (12.8) from Corollary 12.22. For this, let

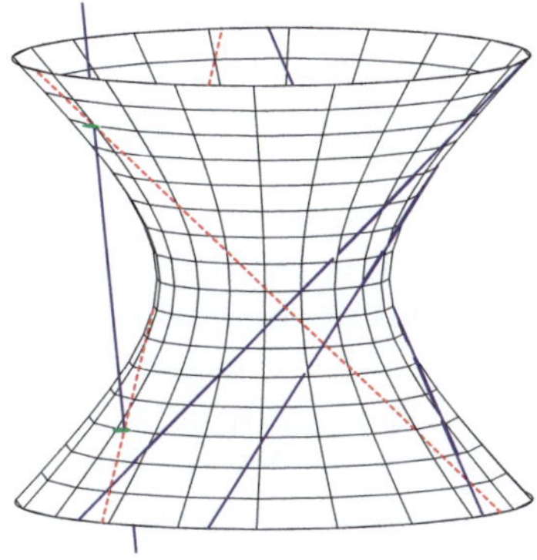

Fig. 12.1 The geometry of the common transversal of four *given lines*. The two transversals are *dashed*

$p_{ij} = x_i y_j - x_j y_i$, and eliminate all x and y variables from the ideal generated by these equations.

```
> ring R = 0, (x1,x2,x3,x4,y1,y2,y3,y4,
               p12,p13,p14,p23,p24,p34), lp;
> ideal I = p12 - (x1*y2 - x2*y1), p13 - (x1*y3 - x3*y1),
            p14 - (x1*y4 - x4*y1), p23 - (x2*y3 - x3*y2),
            p24 - (x2*y4 - x4*y2), p34 - (x3*y4 - x4*y3);
> eliminate(I,x1*x2*x3*x4*y1*y2*y3*y4);
 _[1]=p12*p34-p13*p24+p14*p23
```

The lexicographic Gröbner basis of the ideal I consists of 17 polynomials. By Theorem 10.1, one of these polynomials is the polynomial of Plücker coordinates obtained by the elimination of all x and y variables.

12.5.1 Transversals

Determining all lines ℓ that intersect a given set of lines $\ell_1, \ldots, \ell_k \subseteq \mathbb{R}^3$ is a standard operation in computer graphics. Every line of this type is called a *transversal* of $\ell_1, \ldots, \ell_k$. This problem is well suited to illustrate the passage from linear to non-linear structures. Even though lines are affine subspaces of $\mathbb{R}^3$, there exist—as mentioned in the introduction to this chapter—in general two (possibly complex) transversals for any four given lines.

If $\ell_1, \ldots, \ell_k$ are given in dual Plücker coordinates, then the intersection condition $\ell \cap \ell_i \neq 0$ yields, by Corollary 12.25, the condition

$$f_i(p_{12}, \ldots, p_{34}) = 0,$$

which is linear in the Plücker coordinates p which we want to compute. If $k = 4$, and these conditions are linearly independent, this homogeneous system of equations in $\mathbb{R}^6$ has a 2-dimensional solution space. If v and w are the generators of this solution space, then substituting the general solution $\lambda v + \mu w$ (for $\lambda, \mu \in \mathbb{R}$) into the Plücker equation yields a homogeneous quadratic equation in λ, μ. The solutions can then be easily obtained by dehomogenization.

Actually, this situation has a very nice geometric interpretation. If ℓ_1, ℓ_2 and ℓ_3 are skew, then ℓ_1, ℓ_2 and ℓ_3 either lie in a uniquely determined hyperboloid of one sheet, or in a hyperbolic paraboloid; see Exercises 12.26 and 12.27. In both cases, this quadric contains two families of lines, and ℓ_1, ℓ_2 and ℓ_3 are contained in the same family. In general, ℓ_4 intersects the quadric in two points. The two lines of the other family of lines determined by these two intersections intersect ℓ_1, ℓ_2, ℓ_3 and ℓ_4. See Fig. 12.1 for the case where the first three lines are contained in a hyperboloid of one sheet.

The degenerate cases can also be solved using this approach. If ℓ_4, like the first three lines, is contained in the family of lines defined by the quadric, then each line of the other family of lines intersects the four given lines.

12.6 Exercises

Exercise 12.26 Assume we are given a *hyperboloid H* of the form

$$\frac{x^2}{a^2} + \frac{y^2}{b^2} - \frac{z^2}{c^2} = 1, \quad \text{with } a, b, c > 0.$$

Determine a parametrization of the two families of lines contained in H.

Exercise 12.27 Show that three given pairwise skew lines in $\mathbb{R}^3$ lie on a uniquely determined quadratic hypersurface; specifically, on a hyperboloid of one sheet, or a hyperbolic paraboloid.

Exercise 12.28 Write a `Singular` program that computes for three given skew lines the quadric from the last exercise.

Exercise 12.29 The Plücker coordinates of the set of tangential hyperplanes to the unit sphere $\mathbb{S}^2 \subseteq \mathbb{R}^3$ centered at the origin defines a hypersurface in $\mathbb{P}^5$. What is its defining polynomial?

12.7 Remarks

Plücker coordinates can be traced back to Julius Plücker (1808–1868). Further information on Plücker coordinates and Grassmann manifolds can be found in the classical book of Hodge and Pedoe [63], Pottmann and Wallner [83] and Fischer and Piontkowski [40].

The fact that the exterior multiplication defined on the canonical basis vectors has a unique bilinear extension to the exterior algebra $\bigwedge V$ (as was to be shown in Exercise 12.6) is based on the corresponding universal property of the tensor algebra of V, since $\bigwedge V$ can be written as a quotient of V. See, for example, the treatment in Roman's book [89].

For some contemporary developments in computational line geometry, see the survey article by Sottile and Theobald [92].

Chapter 13
Applications of Non-linear Computational Geometry

In this concluding chapter, we study some applications of non-linear computational geometry. First, we will study Voronoi diagrams for line segments (instead of points), which leads to non-linear edges. Next, we illustrate how some two- and three-dimensional real world problems (from robotics and satellite geodesy) can be formulated in terms of polynomial equations, and how they can be solved using the methods described in the previous chapters. Note that we will give simplified examples and that our focus is always on demonstrating the modeling of these problems with polynomial equations. Many related questions quickly lead to algorithmic and algebraic topics that are beyond the scope of this book.

13.1 Voronoi Diagrams for Line Segments in the Plane

Let $S = \{s^{(1)}, \dots, s^{(m)}\}$ be a finite set of line segments in $\mathbb{R}^2$. We define, analogously to the ordinary Voronoi diagrams in Chapter 6, the *Voronoi region* of $s^{(i)}$ as

$$\mathrm{VR}\big(s^{(i)}\big) := \big\{x \in \mathbb{R}^2 : \mathrm{dist}\big(x, s^{(i)}\big) \leq \mathrm{dist}\big(x, s^{(j)}\big) \text{ for all } 1 \leq j \leq m\big\},$$

where $\mathrm{dist}(x, s^{(i)})$ denotes the Euclidean distance from the point x to the segment $s^{(i)}$. Our first observation is that the Voronoi regions of S are in general *not* polyhedral, but can be described by non-linear arcs (see Fig. 13.1).

Let y be a point of a Voronoi region $\mathrm{VR}(s^{(i)})$ and let z be the point on $s^{(i)}$ which has the shortest distance to y. The segment $[y, z]$ is therefore contained in $\mathrm{VR}(s^{(i)})$. This implies that there exists a convex set (the segment $s^{(i)}$), such that for each point of the Voronoi region of $s^{(i)}$ at least one point of the convex set is visible. Any object with this property is called *weakly star shaped*. With regard to topology, this implies that every Voronoi cell $s^{(i)}$ is (simply) connected.

We first consider the case where the segments $s^{(i)}$ are pairwise disjoint. For given indices $i \neq j$ we study the *bisector curve* (for short: the *bisector*)

$$B_{ij} := \mathrm{VR}\big(s^{(i)}\big) \cap \mathrm{VR}\big(s^{(j)}\big).$$

M. Joswig, T. Theobald, *Polyhedral and Algebraic Methods in Computational Geometry*, 209
Universitext, DOI 10.1007/978-1-4471-4817-3_13,
© Springer-Verlag London 2013

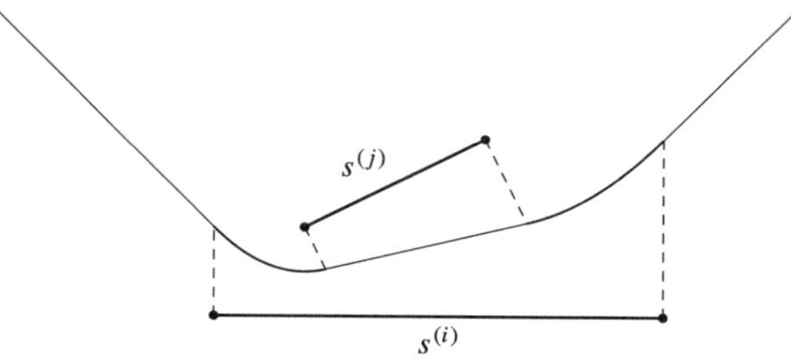

Fig. 13.1 In this example, the bisector of two disjoint segments $s^{(i)}$ and $s^{(j)}$ consists of two parabolic arcs, two infinite rays and a line segment

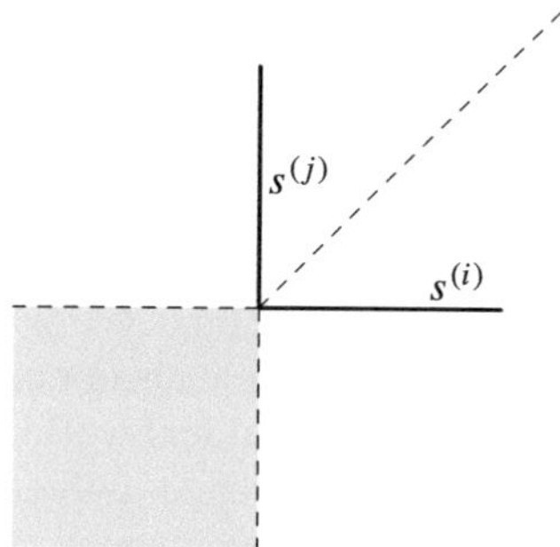

Fig. 13.2 The bisector of two segments $s^{(i)}$ and $s^{(j)}$ that intersect in an endpoint consists of the *shaded region* as well as the *dashed edges*

The bisector B_{ij} is an unbounded and piecewise algebraic curve in the plane (see Fig. 13.1). We saw in Section 6.4 that the set of points which are equidistant from a given point and a given line define a parabola. Therefore, B_{ij} consists of line segments and parabolic arcs. In fact, we have:

Exercise 13.1

(a) The bisector curve B_{ij} of two disjoint segments $s^{(i)}$ and $s^{(j)}$ is an unbounded and piecewise algebraic curve in the plane consisting of at most 7 (possibly unbounded) line segments and parabolic arcs.

(b) The bound 7 is sharp, i.e., there exist pairs of segments whose bisectors consist of exactly 7 line segments and parabolic arcs.

If the two segments $s^{(i)}$ and $s^{(j)}$ share an endpoint, then the bisector B_{ij} is no longer a curve, but rather a two-dimensional set (see Fig. 13.2). Such "2-dimensional Voronoi edges" do not behave nicely and we will not delve deeper into this case.

Instead, we return our focus to disjoint segments. Even though the Voronoi regions do not define a polyhedral complex, the 0-, 1- and 2-dimensional (non-linear) cells define a *cellular decomposition* of $\mathbb{R}^2$. The 1-cells are the linear and parabolic pieces of the bisectors; the 0-cells are the points in between. The non-linear cell-complex constructed in this way is called the *Voronoi diagram* of S. As in the polyhedral case, inclusion defines a partial order on the cells, the f-vector (f_0, f_1, f_2) counts the cells of different dimensions, and we have the two-dimensional version of

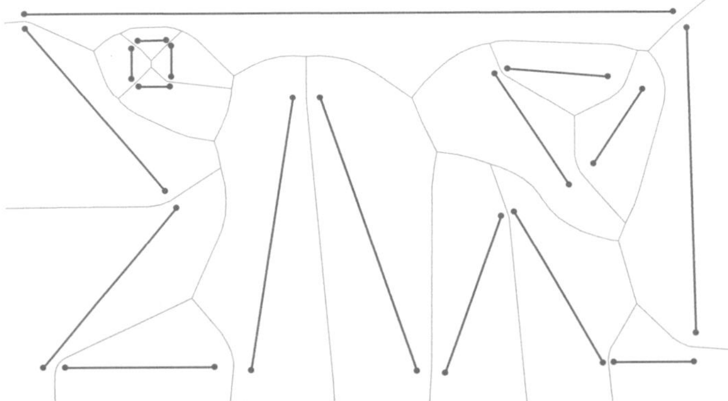

Fig. 13.3 The Voronoi diagram of disjoint line segments

Euler's formula $f_0 - f_1 + f_2 = 1$. We define the *complexity* of the Voronoi diagram for line segments as the sum $f_0 + f_1 + f_2$.

Theorem 13.2 *A Voronoi diagram for m segments has linear complexity $O(m)$.*

Proof The Voronoi diagram of the m segments consists of m connected Voronoi regions. If the segments are in general position, each vertex is contained in exactly three regions. Exercise 13.1 shows that each edge consists of at most 7 line segments and parabolic arcs. By Euler's formula, the complexity of the Voronoi diagram is linearly bounded.

If the segments are not in general position, they can be transformed into general position by perturbation. Through this process, the number of vertices, edges and 2-dimensional faces is not reduced. $\qquad\square$

The beach line algorithm from Section 6.4 can be generalized to a sweep line algorithm to construct the Voronoi diagram of line segments.

Exercise 13.3 Show that the Voronoi diagram of m line segments can be constructed in $O(m \log m)$ steps using the sweep line algorithm.

In Chapter 11 we defined the medial axis of a plane curve C as the topological closure of the set of those points in the plane whose closest point of C is not uniquely defined. If the set S of line segments is regarded as a non-connected curve $C :=$ $\bigcup_{i=1}^{m} s^{(i)}$, then the medial axis of C consists of the vertices and edges of the Voronoi diagram of S.

The software package CGAL can compute the Voronoi diagrams of disjoint line segments. Figure 13.3 displays a possible output.

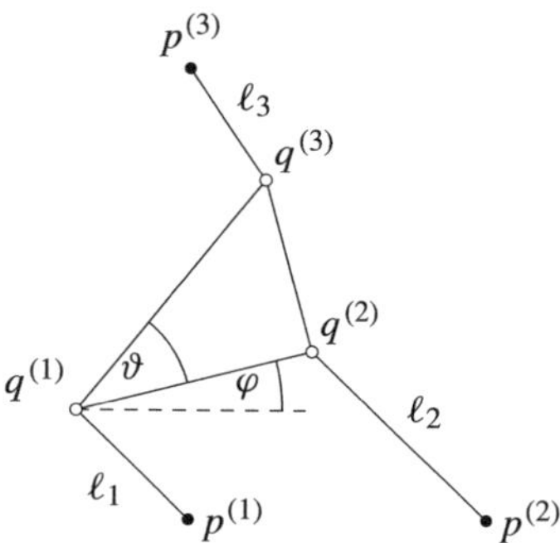

Fig. 13.4 Notations for the direct kinematic problem

13.2 Kinematic Problems and Motion Planning

We now examine elementary robot mechanisms that we will model with a system of rigid elements, joints and axes whose parameters (e.g., the length of an element or the angle between two elements) are variable. In particular, we focus on so-called *manipulators* which are robot mechanisms that are fixed at a certain workspace.

The goal of *kinematics* is to study the geometry and the time dependent aspects of the movement of such mechanisms; the forces causing movement are not taken into consideration.

To begin, we study the following simple robot mechanism in the plane: We have three fixed points $p^{(1)}, p^{(2)}, p^{(3)} \in \mathbb{R}^2$ whose coordinates can be chosen without loss of generality such that $p^{(1)} = (0, 0)$ and $p^{(2)} = (p_{21}, 0)$. Consider the rigid triangle $\triangle$ with vertices $q^{(1)}, q^{(2)}, q^{(3)}$ that is connected to the fixed points via three segments with variable length. We denote the length of the i-th connecting segment by ℓ_i, and assume that there is a freely moving joint at the endpoints of each segment (see Fig. 13.4).

For robot mechanisms it is typically much easier to determine the length of the connecting segments than the Cartesian coordinates of the mobile points. In general, however, the lengths of the connecting segments do not uniquely define the positions of the relevant vertices (the vertices of the triangle $\triangle$ in this case). The so-called *direct kinematic problem* poses the question of determining all possible positions of the triangle for a given set of lengths.

A straightforward (but, as we see below, not optimal) way to model the planar problem is with the following system of equations:

$$
\begin{aligned}
\ell_i^2 &= \left(q^{(i)} - p^{(i)}\right)^2, \quad 1 \le i \le 3, \\
s_{ij}^2 &= \left(q^{(i)} - q^{(j)}\right)^2, \quad 1 \le i < j \le 3
\end{aligned}
\tag{13.1}
$$

where $s_{ij} = \mathrm{dist}(q^{(i)}, q^{(j)})$ denotes for $1 \le i \le 3$ the given distance between $q^{(i)}$ and $q^{(j)}$. We will see below that given generic values for $p^{(i)}$, this system of equations has 12 (complex) solutions for $q^{(1)}, q^{(2)}, q^{(3)}$. These equations (13.1) determine the triangle only up to congruence. Therefore, some of the solutions correspond to reflections of the mobile triangle illustrated in Fig. 13.4. We will later return to this system and its additional solutions. Before we do that, we first study another formulation which avoids these additional unwanted solutions.

If we write $q^{(1)} = (x, y)$, then according to Fig. 13.4, the problem can be modeled by the following system of equations.

$$
\begin{aligned}
\ell_1^2 &= x^2 + y^2, \\
\ell_2^2 &= (x + s_{12}\cos\phi - p_{21})^2 + (y + s_{12}\sin\phi)^2, \\
\ell_3^2 &= \left(x + s_{13}\sin(\phi + \theta) - p_{31}\right)^2 + \left(y + s_{13}\sin(\phi + \theta) - p_{32}\right)^2.
\end{aligned}
\tag{13.2}
$$

Here θ and ϕ denote the angle in the triangle at vertex $q^{(1)}$ and the angle between the triangle and the horizontal axis at point $q^{(1)}$. The solutions to these three equations for the unknowns (x, y, ϕ) are the solutions to the direct kinematic problem.

Since the system of (13.2) contains trigonometric expressions, we need to transform it into a system of polynomial expressions before we can begin using algebraic methods. This can be done by first writing the equations in the form

$$
\begin{aligned}
\ell_1^2 &= x^2 + y^2, \\
\ell_2^2 &= x^2 + y^2 + Rx + Sy + Q, \\
\ell_3^2 &= x^2 + y^2 + Ux + Vy + W
\end{aligned}
$$

where

$$
\begin{aligned}
R &= 2s_{12}\cos\phi - 2p_{21}, \\
S &= 2s_{12}\sin\phi, \\
Q &= -2s_{12}p_{21}\cos\phi + s_{12}^2 + p_{21}^2, \\
U &= 2s_{13}\cos(\phi + \theta) - 2p_{31}, \\
V &= 2s_{13}\sin(\phi + \theta) - 2p_{32}, \\
W &= -2s_{13}\cos(\phi + \theta)p_{31} - 2s_{13}\sin(\phi + \theta)p_{32} + s_{13}^2 + p_{31}^2 + p_{32}^2.
\end{aligned}
$$

To express the trigonometric functions in terms of polynomials, we use the substitutions

$$
\sin\phi = \frac{2T}{1 + T^2} \quad \text{and} \quad \cos\phi = \frac{1 - T^2}{1 + T^2}.
$$

By Lemma 7.1 we know that the stereographic projection

$$
T \mapsto \left(\frac{1 - T^2}{1 + T^2}, \frac{2T}{1 + T^2} \right)
$$

maps the real axis bijectively to $\mathbb{S}^1 \setminus \{(-1, 0)\} \subseteq \mathbb{R}^2$.

We now examine the following example: $p^{(1)} = (0, 0)$, $p^{(2)} = (16, 0)$, $p^{(3)} = (0, 10)$, $s_{12} = 17$, $s_{13} = 21$, $l_1 = 15$, $l_2 = 15$, $l_3 = 12$, and $\sin\theta = 3/5$, where $0 \leq \theta \leq \pi/2$. In Maple this can be expressed as:

```
> with(Groebner):
```

```
> p21 := 16: p31 := 0: p32 := 10:
> s12 := 17: s13 := 21:
> l1 := 15: l2 := 15: l3 := 12:
> sth := 3/5: cth := sqrt(1-sth^2):
```

For the transformation we use the addition theorems

$$\sin(\phi + \theta) = \sin\phi\cos\theta + \cos\phi\sin\theta \ \text{ and } \ \cos(\phi + \theta) = \cos\phi\cos\theta - \sin\phi\sin\theta.$$

So the necessary equations are:

```
> sphi := 2*T/(1+T^2): cphi := (1-T^2)/(1+T^2):
> sphith := sphi*cth + cphi*sth: cphith
            := cphi*cth - sphi*sth:
```

```
> R := 2*s12*cphi - 2*p21:
> S := 2*s12*sphi:
> Q := -2*s12*p21*cphi + s12^2 + p21^2:
> U := 2*s13*cphith - 2*p31:
> V := 2*s13*sphith - 2*p32:
> W := -2*s13*p31*cosphith - 2*s13*p32*sphith
           + s13^2 + p31^2 + p32^2:
```

```
> eq1 := x^2 + y^2 - l1^2;
> eq2 := x^2 + y^2 + R*x + S*y + Q - l2^2;
> eq3 := x^2 + y^2 + U*x + V*y + W - l3^2;
```

The term eq1 is already a polynomial and we have:

```
  eq1 := x^2+y^2-225
```

After applying the addition theorems, the expressions eq2 and eq3 become polynomials in x and y, but only rational functions in T. Multiplying by $1 + T^2$ resolves this and results in polynomials in the unknowns T, x, y:

```
> eq2b := simplify((1+T^2)*eq2);
> eq3b := simplify((1+T^2)*eq3);
```

```
              2      2  2      2      2  2
  eq2b := x  + x  T  + y  + y  T  + 2 x


                2                              2
        - 66 x T  + 68 T y - 224 + 864 T
```

```
              2      2  2      2      2  2                                    2
  eq3b := x   + x   T + y   + y   T + 168/5 x - 168/5 x  T


                                         2
        - 252/5 x T + 336/5 T y + 26/5 y  - 226/5 y T


                            2
        + 145 + 649 T  - 672 T
```

Using the results of previous chapters, we can compute the x-coordinate of the point (x, y) by computing the univariate polynomial in the ideal generated by `eq1`, `eq2b` and `eq3b`. From this we obtain the y-coordinate by continuation of the partial solutions, and via T we obtain the angle ϕ.

```
> p := UnivariatePolynomial(x, [eq1, eq2b, eq3b], {T,x,y});
> xi := fsolve(p,x);
```

The result of this computation is:

```
  p := 4293662653017426246255 + 1783141486293101799200 x


                             2                              3
     + 14077640037857031888 x  - 15259320794001132800 x


                             4                        5
     - 175035877377261312 x  + 3063379125125120 x


                        6
     + 476463824896000 x

  xi := -12.85759683, -9.949639909, -8.770015468, -3.829357647,
        14.07828959, 14.89891479
```

For our particular example, all six solutions are real, however this is not always the case. Figure 13.5 depicts the six solutions of the direct kinematic problem.

We now study a more complicated three-dimensional manipulator, the so-called *Stewart platform*. This manipulator is a robot mechanism which has six points, $p^{(1)}, \dots, p^{(6)}$, fixed in space (usually in the base plane), and six points, $q^{(1)}, \dots, q^{(6)}$, positioned on a rigid body K, which is mobile in space (via translation and rotation). For each i, the points $p^{(i)}$ and $q^{(i)}$ are connected via segments ("legs") of variable length. These legs are connected to the endpoints $p^{(i)}$ and $q^{(i)}$ by ball joints (see Fig. 13.6). Mechanisms of this kind are used in special vehicles and flight simulators.

In the direct kinematic problem for the Stewart platform we want to determine the position and orientation of K for given lengths of the six connecting segments. For each leg the distance condition is defined by an equation. In the modeling process it is common to first choose different coordinate systems Σ_1 and Σ_2 for the basis points and the points on the platform respectively. Let $p^{(i)}$ denote the basis

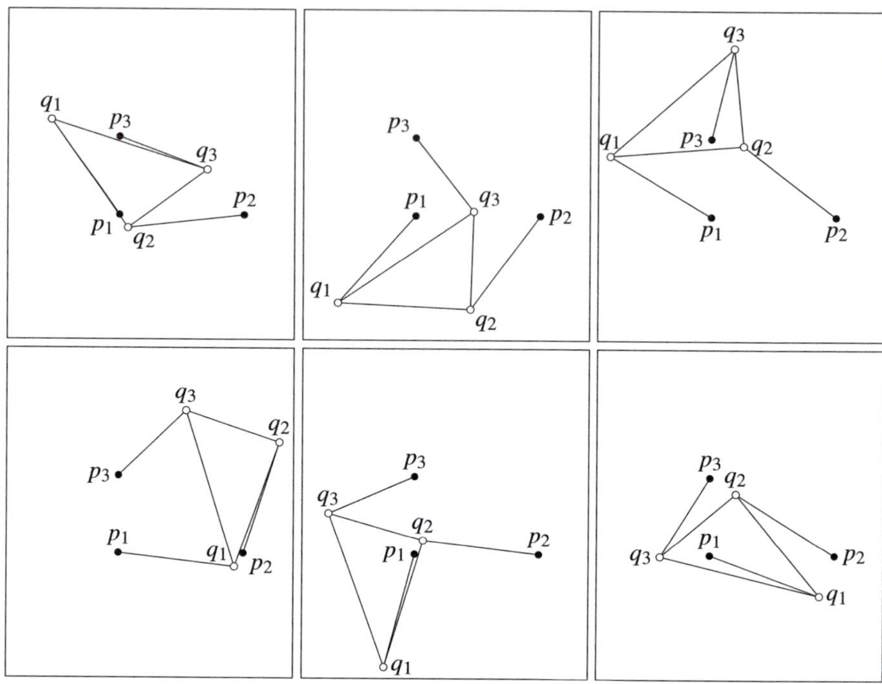

Fig. 13.5 The six solutions of the direct kinematic problem

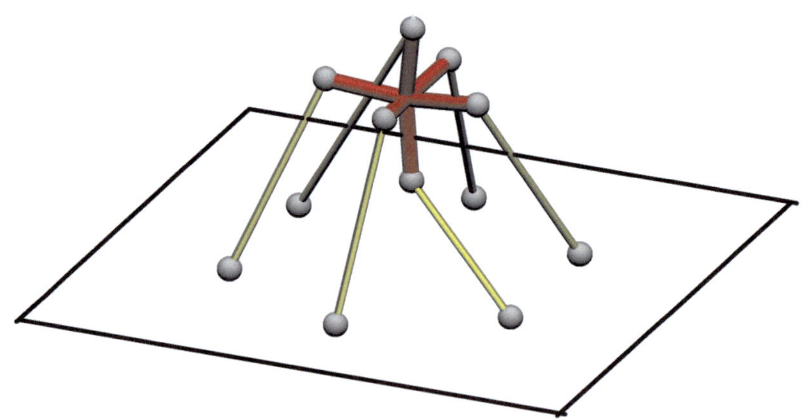

Fig. 13.6 The Stewart platform

points and $q^{(j)}$ denote the points on the platform *with respect to their correspond-ing coordinate systems*. Furthermore, let $x = (x_1, x_2, x_3)$ denote the coordinates of the origin of Σ_2 in Σ_1, and let R denote the orthogonal 3×3-matrix that describes the orientation (i.e., the rotation) of K in the outer coordinate system Σ_1. The equation for the i-th leg can now be written as

$$\left(x + Rq^{(i)} - p^{(i)}\right)\left(x + Rq^{(i)} - p^{(i)}\right) = \ell_i^2. \tag{13.3}$$

The matrix R can be expressed as

$$R = \begin{pmatrix} \cos\alpha\cos\beta & \cos\alpha\sin\beta\sin\gamma - \sin\alpha\cos\gamma & \cos\alpha\sin\beta\cos\gamma + \sin\alpha\sin\gamma \\ \sin\alpha\cos\beta & \sin\alpha\sin\beta\sin\gamma + \cos\alpha\cos\gamma & \sin\alpha\sin\beta\cos\gamma - \cos\alpha\sin\gamma \\ -\sin\beta & \cos\alpha\sin\gamma & \cos\beta\cos\gamma \end{pmatrix}.$$

Substituting the matrix R in (13.3) results in a system of six equations in the six unknowns $x = (x_1, x_2, x_3)$, α, β and γ. To transform this system into a system of polynomial equations we let

$$x_4 = \sin\alpha, \qquad x_5 = \cos\alpha,$$
$$x_6 = \sin\beta, \qquad x_6 = \cos\beta,$$
$$x_7 = \sin\gamma, \qquad x_7 = \cos\gamma$$

and employ the relations

$$x_4^2 + x_5^2 = 1, \qquad x_6^2 + x_7^2 = 1, \qquad x_8^2 + x_9^2 = 1.$$

Combining this with the six equations (13.3) for the legs, we obtain a system of nine equations in nine unknowns.

The direct kinematic problem for the Stewart platform has 40 solutions over $\mathbb{C}$ if the lengths are chosen generically, and there exists lengths such that all 40 solutions are real.

In the following, we focus on a special case, where the points $p^{(i)}$ and $p^{(3+i)}$ lie above one another on a line which is perpendicular to the plane, as if they were placed along a vertical pillar. We also assume that $q^{(i)} = q^{(3+i)}$ for $1 \leq i \leq 3$. This *special Stewart platform* is illustrated in Fig. 13.7. For given lengths ℓ_i of the connecting segments the possible endpoints of the segments $[p^{(1)}, q^{(1)}]$ and $[p^{(4)}, q^{(4)}]$ define a circle C_1 in a horizontal plane of $\mathbb{R}^3$. The center of this circle lies on the line connecting $p^{(1)}$ and $p^{(4)}$. Similar statements hold for the remaining pairs of the connecting segments. Hence, we can replace each of the connected pairs by just one connection that rotates around the corresponding vertical axis $\mathrm{aff}\{p^{(i)}, p^{(3+i)}\}$ (see Fig. 13.7). The radii of the circles C_1, C_2 and C_3 are denoted by r_1, r_2 and r_3.

Let H_i be the plane that contains the circle C_1. For each $1 \leq i \leq 3$, the plane H_i is parallel to the base plane. Consider the orthogonal projections $\pi(C_2)$ and $\pi(C_3)$ on H_1. Every movement of $q^{(2)}$ along C_2 induces a movement of $\pi(q^{(2)})$ along the circle $\pi(C_2)$. The length of the edge $[q^{(1)}, q^{(2)}]$ of the triangle $\mathrm{conv}\{q^{(1)}, q^{(2)}, \pi(q^{(2)})\}$ is constant; the length of the edge $[q^{(2)}, \pi(q^{(2)})]$ is constant as well and equals the distance between the planes H_1 and H_2. Since the angle $(q^{(2)}, \pi(q^{(2)}), q^{(1)})$ is a right angle, the distance of $\pi(q^{(2)})$ to $q^{(1)}$ is constant. There are corresponding statements for the triangle $\mathrm{conv}\{q^{(1)}, q^{(3)}, \pi(q^{(3)})\}$. Hence, we obtain a triangle $\mathrm{conv}\{q^{(1)}, \pi(q^{(2)}), \pi(q^{(3)})\}$ in the plane H_1 with constant edge

Fig. 13.7 The special
Stewart platform

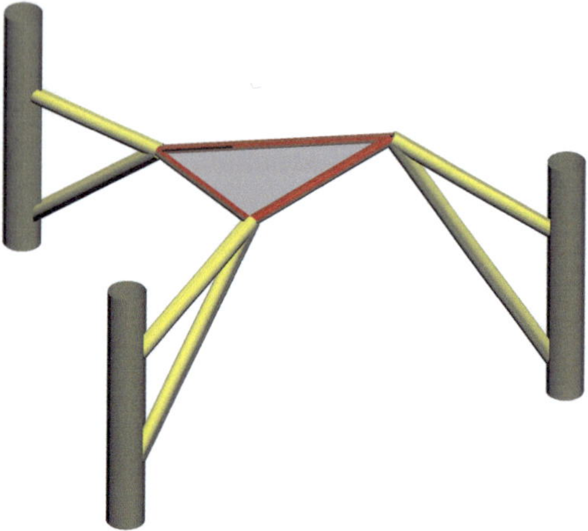

lengths and vertices that are connected to fixed points by segments with lengths
r_1, r_2 and r_3. This is the exact situation of the planar robot mechanism that we studied earlier, but here we cannot exclude the solutions coming from reflections. For every triangle that satisfies the distance requirements, there exists a reflection that satisfies the conditions. We formally state these results with a corollary.

Corollary 13.4 *A special Stewart platform (in general position) has exactly twelve real solutions if the corresponding planar robot mechanism has twelve real solutions (counting reflections).*

We can now see how this problem can be formulated in $\mathtt{Singular}$. Here, the variables p1p, p2p, p3p denote the orthogonal projections of $p^{(1)}$, $p^{(2)}$, $p^{(3)}$ on the plane H_1.

```
> LIB "solve.lib";
> ring R = 0, (q11,q12,q21,q22,q31,q32), (lp);

> vector p1p, p2p, p3p;
> int s1p,s2p,s3p;
> int r1, r2, r3;
> int q13, q23, q33;

> p1p = [1,2]; p2p = [7,4]; p3p = [4,5];
> s1p = 3; s2p = 5; s3p = 7;
> r1 = 7; r2 = 8; r3 = 9;
> q13 = 0; q23 = 0; q33 = 0;

> poly f1 = (q11-q21)^2 + (q12-q22)^2 + (q13-q23)^2 - s1p^2;
> poly f2 = (q11-q31)^2 + (q12-q32)^2 + (q13-q33)^2 - s2p^2;
> poly f3 = (q21-q31)^2 + (q22-q32)^2 + (q23-q33)^2 - s3p^2;
> poly f4 = (q11-p1p[1])^2 + (q12-p1p[2])^2 - r1^2;
```

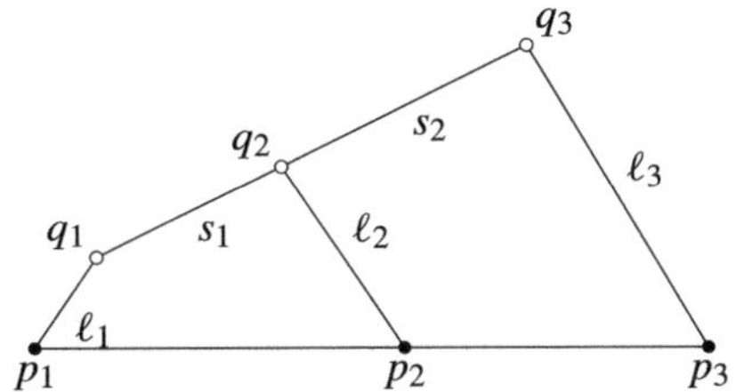

Fig. 13.8 A degenerate planar mechanism


```
> poly f5 = (q21-p2p[1])^2 + (q22-p2p[2])^2 - r2^2;
> poly f6 = (q31-p3p[1])^2 + (q32-p3p[2])^2 - r3^2;

> ideal I = f1, f2, f3, f4, f5, f6;
```

We obtain the q_{11}-coordinates of the mobile triangle via

```
> ideal J = eliminate(I, q12*q21*q22*q31*q32);
> laguerre_solve(J[1]);
```

Finally, we have twelve complex solutions, of which eight are real:

```
[1]:  -2.8003292 , [2]: -2.69678323 ,
[3]:  1.453208 , [4]: 2.18551634 ,
[5]:  2.37409582 , [6]: 4.28762055 ,
[7]:  6.82883162 , [8]: 7.309411 ,
[9]:  (6.39757021+i*1.05308349) ,
[10]: (6.39757021-i*1.05308349) ,
[11]: (7.71054443+i*1.71522046) ,
[12]: (7.71054443-i*1.71522046)
```

Exercise 13.5 Consider the special case of the planar robot mechanism where the fixed points $p^{(1)}$, $p^{(2)}$ and $p^{(3)}$ are collinear and where the triangle with the vertices $q^{(1)}$, $q^{(2)}$ and $q^{(3)}$ degenerates to a segment with sections of length s_1 and s_2 (see Fig. 13.8). How many solutions does the direct kinematic problem for generic lengths ℓ_1, ℓ_2 and ℓ_3 have?

13.3 The Global Positioning System GPS

The *Global Positioning System (GPS)* is a global navigation satellite system. It operates with the help of satellites which continuously orbit the earth in such a way that from almost every point on the earth's surface, at any given time, it is possible to reach at least four different satellites via a straight line. In the early years of GPS there were 18 satellites. Now there are 24. Each satellite continuously emits messages containing the actual position of the satellite and the exact time when the message was sent. There are Earth-based stations that synchronize the satellites' clocks and inform them about their current movement.

Fig. 13.9 The Apollonius problem, a "planar variant" of the GPS problem

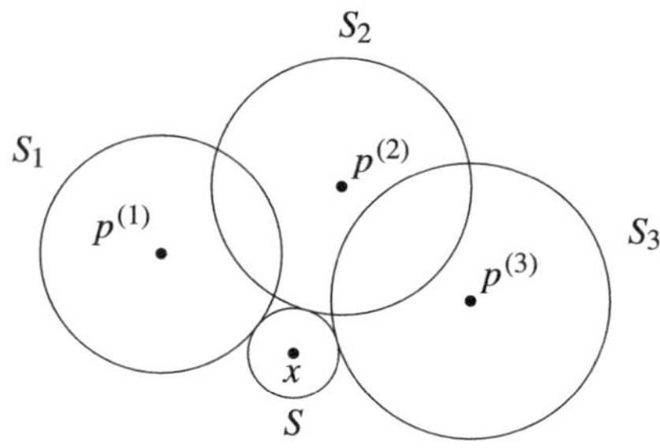

With a small hand-held receiver, we can determine within seconds our actual position within a few meters. To do this, the hand-held device, whose position we denote by x, simultaneously receives signals from at least four satellites with positions $p^{(1)}, \ldots, p^{(4)} \in \mathbb{R}^3$.

The receiver computes the time it took the signal to reach the device and therefore knows its distance to the sender. Since it is impossible to fully synchronize the clocks of the sender and the receiver, the distance can only be computed up to a constant z. If the clock on the receiving side runs slightly too slowly, the measured time difference, and therefore the computed distance, becomes a little smaller. Hence, we speak of determining the *pseudo distances*

$$r_i := \left\| x - p^{(i)} \right\| - z.$$

This leads to the following system of equations:

$$\left(x_1 - p_1^{(i)}\right)^2 + \left(x_2 - p_2^{(i)}\right)^2 + \left(x_3 - p_3^{(i)}\right)^2 = (z + r_i)^2, \quad 1 \le i \le 4. \qquad (13.4)$$

We can think of the pseudo distance r_i as a radius of a sphere S_i with center $p^{(i)}$; the two-dimensional case (with three circles) is illustrated in Fig. 13.9. The center x of the sphere S with radius z has to touch the four spheres $S_1, \ldots, S_4$. The GPS problem is therefore strongly related to the classical geometric Apollonius problem (in its three-dimensional version), which asks for spheres that touch four given spheres in $\mathbb{R}^3$. The position x we were looking for is one of the 16 solutions to the three-dimensional Apollonius problem for the given spheres $S_1, \ldots, S_4$. Since S touches either all four spheres S_i from the outside (if $z > 0$), or all S_i from the inside (if $z < 0$), the system of (13.4) defines only 2 of the 16 solutions of the Apollonius problem. The correct solution is usually the one with the smaller radius r, since the inaccuracy of time is small.

We study the example from Table 13.1 using `Maple`. We assume in the following that the variables `p[i,j]` and `r[i]` are already initialized with values from the table. Using the commands

```
> for i from 1 to 4 do
>    f[i] := (x1-p[i,1])^2 + (x2-p[i,2])^2
              + (x3-p[i,3])^2 - (z+r[i])^2;
> od;
```

Table 13.1 Sample data for the GPS problem, taken from [9]

i	$p_1^{(i)}$	$p_2^{(i)}$	$p_3^{(i)}$	r_i
1	14832308660	−20466715890	−7428634750	24310764064
2	−15799854050	−13301129170	17133838240	22914600784
3	1984818910	−11867672960	23716920130	20628809405
4	−12480273190	−23382560530	3278472680	23422377972

All lengths are given in 10^{-3} m

Table 13.2 Solution to the GPS problem for the data from Table 13.1

x_1	x_2	x_3	z
−2892123412	7568784349	−7209505102	−57479918164.14
1111590460	−4348258631	4527351820	−100000.55

we generate the system of equations. To obtain all solutions for, say x_1, we compute:

```
> with(Groebner):
> g := UnivariatePolynomial(x1, [f[1],f[2],f[3],f[4]],
                        {x1,x2,x3,z}):
> fsolve(g,x1,complex);
```

Numerically this leads to the solution

```
 -2892123412., 1111590460.
```

The two numerical solutions for (x_1, x_2, x_3, z) can be read from Table 13.2. In this case, the second solution is the correct one.

If more than four satellites can be reached simultaneously, the computations can be performed with a higher precision. This leads to a number of numerical topics and questions about stability.

13.4 Exercises

Exercise 13.6 For $a \geq 0$ and $n \geq 1$ compute the Voronoi diagram of the line segments

$$\text{conv}\left\{(ai, i)^T, ((a+1)i, i)^T\right\}, \quad 1 \leq i \leq n$$

corresponding to a staircase configuration.

Exercise 13.7 Generalize the characterization of the vertices of a Voronoi diagram via Voronoi circles (Corollary 6.15) to Voronoi diagrams of line segments.

Exercise 13.8 Given four spheres with radius $r > 0$ centered at the vertices of a regular tetrahedron in $\mathbb{R}^3$, determine all spheres touching the four given spheres.

13.5 Remarks

A detailed description of Voronoi diagrams for line segments can be found in the monograph of Boissonnat and Yvinec [15].

For further material concerning robot motion planing and kinematic problems see the book by McCarthy [77] and the survey article by Halperin, Kavraki and Latombe [57]. An example of a Stewart platform (which is sometimes referred to as a Stewart Gough platform) with 40 real solutions was described by Dietmaier [36]. The kinematic problem for the special Stewart platform was studied by Lazard and Merlet [75].

Further information about the computational geometric questions concerning the Global Positioning System can be found in the book by Awange and Grafarend [9].

Appendix A
Algebraic Structures

We introduce some fundamental algebraic terms here which can also be found in any regular introduction to the subject, for example in Herstein [59] or Lang [74]. The main purpose is to standardize our notation.

A.1 Groups, Rings, Fields

Definition A.1 A non-empty set G with a binary operation $\circ$ is called a *group* if the following conditions are satisfied:

(a) associativity: $(a \circ b) \circ c = a \circ (b \circ c)$ for all $a, b, c \in G$;
(b) there exists a *neutral element* e, i.e., we have $e \circ a = a \circ e = a$ for all $a \in G$;
(c) every element a has an *inverse*, i.e., there exists an element $b \in G$ such that $a \circ b = b \circ a = e$.

If commutativity holds (i.e., $a \circ b = b \circ a$ for all $a, b \in G$) in addition to the group axioms, we call G *abelian*. A *semi-group* is a non-empty set G with a binary operation $\circ$ satisfying conditions (a) and (b).

Definition A.2 A non-empty set R with two binary operations $+$ and $\cdot$ ("addition" and "multiplication") is called a *ring* if the following hold:

(a) $(R, +)$ is an abelian group with neutral element 0;
(b) $(R, \cdot)$ is a semi-group;
(c) the distributive laws hold: $a(b + c) = ab + ac$ and $(a + b)c = ab + ac$.

A ring is called *commutative* if multiplication is commutative.

An *identity element* $1 \in R \setminus \{0\}$ in a ring is a neutral element with respect to multiplication. All rings that we come across in this text, unless otherwise stated, have an identity element. The set

$$R^\times := \{a \in R : \text{there exists } b \in R \text{ such that } ab = 1\}$$

M. Joswig, T. Theobald, *Polyhedral and Algebraic Methods in Computational Geometry*, 223
Universitext, DOI 10.1007/978-1-4471-4817-3,
© Springer-Verlag London 2013

is a group with respect to multiplication, which is called the *group of units* of R. If $(R \setminus \{0\}, \cdot)$ is an abelian group then $(R, +, \cdot)$ is a *field*.

There exist rings that contain non-zero elements a and b such that $ab = 0$. In this case, a and b are called *zero divisors*. In rings without zero divisors we can cancel, i.e., $ac = bc$ implies $(a - b)c = 0$ and hence $a = b$ if $c \neq 0$. A commutative ring without zero divisors is called an *integral domain*.

Let R be an integral domain (with identity element). An element $p \in R \setminus \{0\}$ with $p \notin R^\times$ is said to be *irreducible* if for any decomposition $p = ab$ with $a, b \in R$ we have that $a \in R^\times$ or $b \in R^\times$. An element $p \in R \setminus \{0\}$ with $p \notin R^\times$ is *prime* if for all $a, b \in R$ such that $p|ab$ it follows that $p|a$ or $p|b$. The ring R is called a *unique factorization domain* if every non-zero element that is not a unit is a prime element or the product of finitely many prime elements.

In unique factorization domains the set of prime elements and the set of irreducible elements are the same. Furthermore, the decomposition of an element into its prime factors is unique up to units and ordering. More precisely: If $a \in R \setminus (\{0\} \cup R^\times)$ has the prime factor decomposition $a = p_1 \cdot \ldots \cdot p_r = q_1 \cdot \ldots \cdot q_r$, then $r = s$ and, after a suitable permutation of the q_i, we have $p_i = e_i q_i$ with unit elements e_i for $i \in \{1, \ldots, r\}$.

Example A.3 The ring

$$\mathbb{Z}[\sqrt{-5}] = \{a + b\sqrt{-5} : a, b \in \mathbb{Z}\}$$

is not a unique factorization domain. The number 6 has the decompositions

$$6 = 2 \cdot 3 = (1 + \sqrt{-5})(1 - \sqrt{-5}),$$

and we can show that all factors 2, 3, $1 + \sqrt{-5}$, $1 - \sqrt{-5}$ that appear are irreducible elements of $\mathbb{Z}[\sqrt{-5}]$. Moreover, 1 and -1 are the only units and thus the two factorizations of 6 are truly distinct.

By analogy with the definitions of integers and rational numbers, we can define for an integral domain R the *quotient field Q* of R. The elements of Q are the "fractions" p/q where $p \in R$ and $q \in R \setminus \{0\}$. Addition and multiplication in Q are defined just as the corresponding operations for rational numbers:

$$\frac{p}{q} + \frac{s}{t} = \frac{pt + qs}{qt} \quad \text{and} \quad \frac{p}{q} \cdot \frac{s}{t} = \frac{ps}{qt}.$$

Two elements $\frac{p}{q}$ and $\frac{p'}{q'}$ represent the same element of Q if and only if $pq' = p'q$.

A.2 Polynomial Rings

Let R be a commutative ring with an identity element. Then the set of all (formal) polynomials $a_n x^n + \cdots + a_1 x + a_0$ with $a_i \in R$ in the unknown x defines a ring.

Addition and multiplication of two polynomials $f = \sum_{i=0}^{n} a_i x_i$ and $g = \sum_{j=0}^{m} b_j x^j$ are defined via

$$f + g := \sum_{i=0}^{\max(m,n)} (a_i + b_i)x^i,$$

$$f \cdot g := \sum_{i=0}^{m+n} c_i x^i \quad \text{where } c_i := \sum_{j+k=i} a_j b_k.$$

Here we agree to write $a_i = b_j = 0$ for all $i > n$ and all $j > m$. The ring of coefficients R is embedded in $R[x]$ via the constant polynomials. A unit in R is also a unit in $R[x]$. For integral domains R we have $R[x]^\times = R^\times$. We say that $R[x]$ is generated from R by adjoining the unknown x.

Over a finite field K there exist several polynomials whose corresponding functions

$$K \to K \ : \ x \mapsto f(x)$$

are identical; in the case of fields with an infinite number of elements the mapping of a polynomial to its corresponding function is always injective. See Exercise 10.30.

When studying polynomial rings the following statement is absolutely essential:

Theorem A.4 *If R is a unique factorization domain, then $R[x]$ is a unique factorization domain.*

We can deduce inductively that for each unique factorization domain R the ring of polynomials $R[x_1, \ldots, x_n]$ in the unknowns $x_1, \ldots, x_n$ is also a unique factorization domain.

For a field K, the quotient field of the polynomial ring $K[x_1, \ldots, x_n]$ is called the *field of rational functions* over K which is usually denoted by $K(x_1, \ldots, x_n)$.

A field K is *algebraically closed* if every non-constant polynomial f in $K[x]$ has a root in K, i.e., an element $a \in K$ with $f(a) = 0$. We have:

Theorem A.5 *Every algebraically closed field has an infinite number of elements.*

Idea of proof Assume that a field K has only finitely many elements $a_1, \ldots, a_k$. Then we can use a Lagrange interpolation polynomial to construct a polynomial f of degree $k - 1$ such that $f(a_i) = 1$ for all i. $\qquad\square$

For every field there exists an *algebraic closure*, i.e., an algebraically closed field that contains K and that is minimal with respect to inclusion. The algebraic closure is unique up to isomorphism.

Appendix B
Separation Theorems

The interplay between analysis and convexity gives rise to a rich theory. For a detailed account we refer to the monograph of Gruber [55]. An introductory approach can also be found in Grünbaum [56, § 2].

Two sets $A, B \subseteq \mathbb{R}^n$ are (*strictly*) *separated* if there exists an affine hyperplane H such that $A \subseteq H_{\mathrm{o}}^{+}$ and $B \subseteq H_{\mathrm{o}}^{-}$ (see (2.4) and (2.5)). If A and B are each only in the *closed* affine subspaces of H, then we say that they are *weakly separated*.

A subset of $\mathbb{R}^n$ is called *compact* if it is closed and bounded. Polytopes are compact.

Theorem B.1 *Let C be a closed convex set in $\mathbb{R}^n$ and $p \in \mathbb{R}^n \setminus C$. Then there exists a hyperplane $H \subseteq \mathbb{R}^n$ with $p \in H$ and $H \cap C = \emptyset$.*

Since every convex set is connected, but $\mathbb{R}^n \setminus H$ is not connected, we therefore have that p is weakly separated from C.

Proof Without loss of generality we can assume $p = 0$ and $C \neq \emptyset$. Let c be an arbitrary point of C and let $\bar{B} := \bar{B}(0, \|c\|)$ be the closed ball with center 0 and radius $\|c\|$, where $\| \cdot \|$ denotes the Euclidean norm.

Since the set $C \cap \bar{B}$ is non-empty and compact, the minimum with respect to the Euclidean norm is attained on the set $C \cap \bar{B}$ at a point b. Let $H := \{x \in \mathbb{R}^n : \sum_{i=1}^{n} b_i x_i = 0\}$. Since $p \notin C$, we have $b \neq 0$. Also $0 \in H$, so it suffices to show that

$$\langle b, c \rangle = \sum_{i=1}^{n} b_i c_i \geq \|b\|^2 > 0 \tag{B.1}$$

for all $c \in C$.

Assume there exists a point $c \in C$ with $\sum_{i=1}^{n} b_i c_i < \|b\|^2$. Since C is convex, it contains the segment $[b, c]$ and the points of this segment have the form

$$x(\lambda) := b + \lambda(c - b), \quad 0 \leq \lambda \leq 1.$$

M. Joswig, T. Theobald, *Polyhedral and Algebraic Methods in Computational Geometry*,   227
Universitext, DOI 10.1007/978-1-4471-4817-3,
© Springer-Verlag London 2013

We now show that there exists a $\lambda \in (0, 1)$ with $\|x(\lambda)\| < \|b\|$, contradicting the choice of b. For this, consider the differentiable function of λ defined by

$$\phi : \mathbb{R} \to \mathbb{R}, \quad \phi(\lambda) := \|b\|^2 - \|x(\lambda)\|^2 = -\lambda^2 \|c - b\|^2 - 2\lambda \langle b, c - b \rangle.$$

The derivative at $\lambda = 0$ is $2(\|b\|^2 - \langle b, c \rangle) > 0$. Hence, there exists an $\epsilon > 0$ such that $\|x(\lambda)\| = \|b + \lambda(c - b)\| < \|b\|$ for $0 < \lambda < \epsilon$. $\qquad\square$

Examining the proof carefully, one can see that p and C are strictly separated.

Corollary B.2 *Let C be a closed convex set in $\mathbb{R}^n$ and $p \in \mathbb{R}^n \setminus C$. Then there exists a hyperplane $H \subseteq \mathbb{R}^n$ with $p \in H_\circ^-$ and $C \subseteq H_\circ^+$.*

Proof Since the inequality $\langle b, c \rangle > 0$ in (B.1) is strict, we can move the hyperplane H constructed in the proof of Theorem B.1 slightly towards C without touching C. The explicit calculation is analogous to that in the proof of Theorem 3.8, see Fig. 3.3. $\qquad\square$

An affine hyperplane H is called a *supporting hyperplane* for a convex set $C \subseteq \mathbb{R}^n$ if $H \cap C \neq \emptyset$ holds and C is completely contained in one of the closed affine half-spaces H^+ or H^-. Therefore, at least one of the open half-spaces $H_\circ^+$ or $H_\circ^-$ has an empty intersection with C.

In the case $\dim C < n$ it is possible that both open half-spaces have an empty intersection with C; thus for a non-full-dimensional and non-empty C any hyperplane containing C is a supporting hyperplane.

Corollary B.3 *Let C be a closed convex subset of $\mathbb{R}^n$. Then every point of the boundary of C is contained in a supporting hyperplane.*

Proof Without loss of generality let $p = 0$ be a point on the boundary of C. Since p is a boundary point of C there exists a sequence $(p^{(k)})_{k \in \mathbb{N}}$ outside of C that converges to the origin. For each sequence element $p^{(k)}$ there exists by Theorem B.1 a hyperplane

$$H^{(k)} = \left\{ x \in \mathbb{R}^n : b^{(k)} + \sum_{i=1}^{n} a_i^{(k)} x_i = 0 \right\},$$

with $a^{(k)} \in \mathbb{R}^n \setminus \{0\}$ and $b^{(k)} \in \mathbb{R}$, such that C is contained in the half-space

$$\left(H^{(k)}\right)^+ = \left\{ x \in \mathbb{R}^n : b^{(k)} + \sum_{i=1}^{n} a_i^{(k)} x_i \geq 0 \right\}.$$

We can further assume that $\|a^{(k)}\| = 1$. Then $|b^{(k)}|$ is the Euclidean distance from $H^{(k)}$ to the origin. Since $p^{(k)}$ converges to the origin the sequence $(a^{(k)}, b^{(k)})$

is bounded in $\mathbb{R}^{n+1}$. By the Bolzano–Weierstrass Theorem there exists a convergent subsequence (see [86]). Let (a, b) be the limit of that subsequence and $H = \{x \in \mathbb{R}^n : b + \sum_{i=1}^{n} a_i x_i = 0\}$ the hyperplane defined by this point. By continuity it follows that $b = 0$ and that C is contained in the half-space

$$H^+ = \left\{ x \in \mathbb{R}^n : \sum_{i=1}^{n} a_i x_i \geq 0 \right\}.$$

Since $0 \in H$ we have that H is a supporting hyperplane to C. $\square$

The theorems introduced here have numerous specializations and versions which are also usually called *separation theorems* in the literature. Sometimes Farkas' lemma from Exercise 4.26 is included under this label as well.

Appendix C
Algorithms and Complexity

Here we explain some terms regarding algorithms and complexity. Systematic introductions can be found, for example, in the books by Cormen, Leiserson, Rivest and Stein [27] and Garey and Johnson [46].

C.1 Complexity of Algorithms

Usually, the quality of an algorithm is measured by its run-time and the required storage space. The demand of resources is typically measured in relation to the size of the input.

The *coding length* (or *size*) sizeof(x) of a data object x is the number of bits which are necessary to store the object in the computer. The computational model which we employ is the commonly used *Turing machine* (or its real-world counterpart, the *von Neumann computer*). A natural number $n > 0$ has, say, a binary representation with $\lfloor \log_2 n \rfloor + 1$ digits, so we have sizeof(n) = $\lfloor \log_2 n \rfloor + 1$. Rational numbers can be coded as pairs of natural numbers with an additional bit for their sign. Matrices or polynomials are stored as the sequence of their coefficients (of rational numbers, for example) and so forth.

The *run-time complexity* $t_A(n)$ of an algorithm A denotes the maximal number of steps that A needs to obtain a solution for one instance of the problem of coding length n. Analogously, the *space complexity* $s_A(n)$ denotes the maximal number of storage cells that are needed to solve one instance of the problem of size n. Our focus is on the run-time complexity of algorithms.

Often it is impossible to determine the exact complexity of an algorithm A. Usually we are interested in determining, as accurately as possible, the growth of the functions $t_A(n)$ and $s_A(n)$ with respect to the input size n. Bounds for the growth serve as a measure of the quality of an algorithm.

It is useful to neglect constant factors since we do not want to take technical aspects such as the specific hardware (within our computational model) or the programming language into consideration. Furthermore, it is practical to ignore all non-

M. Joswig, T. Theobald, *Polyhedral and Algebraic Methods in Computational Geometry*,   231
Universitext, DOI 10.1007/978-1-4471-4817-3,
© Springer-Verlag London 2013

dominant terms of the complexity functions that occur in the complexity analysis. This is called *asymptotic analysis*.

For the asymptotic characterization of the upper bound of a complexity function $f : \mathbb{N} \to \mathbb{R}_{\geq 0}$ we use the notation

$$f \in O(g)$$

if two constants $c, n_0 \in \mathbb{N}$ exist such that for all $n \geq n_0$

$$f(n) \leq c \cdot g(n).$$

We say that "f is at most of order g". It is also common to use $O(n)$ as a term in arithmetic expressions.

Example C.1 The class $O(1)$ is the class of functions bounded by a constant. $f \in n^{O(1)}$ means that f is bounded above by a polynomial in n.

When studying lower bounds for a complexity function f we use the following notation. We write

$$f \in \Omega(g),$$

which is read as "f is at least of order g" if there exist two constants $c, n_0 \in \mathbb{N}$ such that for all $n \geq n_0$

$$f(n) \geq c \cdot g(n).$$

We write

$$f \in \Theta(g),$$

if $f \in O(g)$ and $g \in O(f)$, i.e., if f and g have the same order of growth.

Example C.2 (Binary Search) Given an increasing sequence $(a_1, \ldots, a_n)$ of pairwise distinct natural numbers and a number $x \in \mathbb{N}$, we want to determine algorithmically if x is contained in the sequence. A naive method would be to compare x successively with every element $a_1, \ldots, a_n$. This method needs $\Theta(n)$ steps in its worst case, which occurs when x is not contained in the sequence.

Since our sequence was given in a monotonic order, the "divide-and-conquer" principle decreases the number of steps required. By comparing x to $a_{\lfloor n/2 \rfloor}$, we can determine whether x is contained in the first or second half of the sequence. By recursively repeating this step, we can determine in $O(\log n)$ many steps whether x is contained in the sequence.

This binary search principle is employed, for example, when determining the closest neighbor in Section 6.5.

The Algorithm 5.4 from Section 5.3 which computes the convex hull in the plane is also based on the "divide-and-conquer" principle.

Algorithm C.1: `MergeSort`

---

1 **Partition.** The sequence A is partitioned into two subsequences
  $A_1 = (a_1, \ldots, a_{n/2})$, $A_2 = (a_{n/2+1}, \ldots, a_n)$.
2 **Recursion.** Every subsequence is sorted recursively using the same method.
  Let B_1 and B_2 be the two resulting sorted subsequences.
3 **Merge.** Merge the two sorted sequences B_1 and B_2 into a new sorted sequence
  for the sequence A.

A simple example which illustrates several paradigms of efficient algorithms is
the problem of sorting numbers. This problem is also essential for many geometric
algorithms (i.e., for planar convex hull algorithms). We have:

Theorem C.3 *Sorting n numbers can be done in $O(n \log n)$ steps.*

Proof (Sketch)We consider, without loss of generality, a sequence $A = (a_1, \ldots, a_n)$
of pairwise disjoint numbers, where n is a power of 2. In the following we illustrate
the algorithm *merge sort*, that is also based on the "divide-and-conquer" principle
and is a method that does not exceed the upper bound for the run-time. The Algorithm C.1 consists of the three steps described below.

We have the following recursive relation for the run-time $t(n)$ of merge sort

$$t(n) \leq 2t\left(\frac{n}{2}\right) + dn$$

where $d > 0$ is a constant. Solving this recursion yields the upper bound for sorting
algorithms. $\square$

A fundamental statement of complexity theory says that no algorithm based on
the comparison of numbers as its elementary step can have an asymptotically better
run-time than merge sort. This can be proved using a decision tree model, see [27].
Through this we obtain an asymptotically exact estimation of the run-time complexity of the sorting problem.

Theorem C.4 *Sorting n numbers based on comparisons has complexity $\Theta(n \log n)$.*

C.2 The Complexity Classes P and NP

A *decision problem* is an algorithmic problem that has only two possible solutions:
"Yes" or "No". An *optimization problem* requires finding an *optimal* solution from a
possibly large set of feasible solutions. The quality of a solution is measured using a
cost function. Every optimization problem induces a filtration of decision problems:

The optimization problem $\max\{c(x) : x \in X\}$ and the bound k suggest the question of whether there exists a solution $x \in X$ such that $c(x) \geq k$.

Determining if a class of algorithms is considered *efficient* depends on the specific application. In the context of optimization, only those algorithms whose run-time is bounded above by a polynomial expression in the coding length of the input are considered efficient. In general, we try to avoid algorithms with exponential costs. In contrast, we have that for Gröbner bases, as seen in Chapter 9, the currently known algorithms have a run-time complexity which is *doubly exponential* in the input length. Despite this, many modern applications rely on such methods.

Definition C.5 An algorithm A is called a *polynomial time algorithm* if there exists a univariate polynomial p such that for every input x, the algorithm A terminates in $O(p(\mathrm{sizeof}(x)))$ steps.

An important goal of complexity theory is to determine which problems have such algorithms. In the following we will focus mainly on decision problems. As a measurement for *efficiency* we use Definition C.5.

The Complexity Class P   The class P (polynomial time) denotes the set of all decision problems for which there exists a polynomial time algorithm that solves the problem.

The class of algorithms which only need a polynomially bounded storage space is called PSPACE. Clearly, P is contained in PSPACE.

The Complexity Class NP   The class NP (non-deterministic polynomial time), which we define below, consists of those problems which have an efficient non-deterministic solution algorithm. In contrast to the deterministic case, where there is exactly one possible step at each stage, the non-deterministic approach allows various possible actions at each stage.

Consider, for example, the search for a proof of a mathematical theorem. If the statement is wrong, there exists no such proof, but if a proof for the statement exists, then there is often more then one proof. To show that the theorem is correct it is of course sufficient to show that *at least* one proof exists. Finding a proof can be arbitrarily difficult. Once we are given a proof, it is in general not as difficult to verify the correctness of the proof and thus accept the theorem. In complexity theory such proofs are also referred to as *certificates* (or *witnesses*).

Definition C.6 A decision problem $\mathcal{A}$ is contained in NP if there exists a polynomial p and a polynomial algorithm A such that for every input x and every possible certificate y of size at most $p(\mathrm{sizeof}(x))$, the algorithm A computes a value $t(x, y)$ that satisfies the following:

(a) If the answer to the input x is "No" then $t(x, y) = 0$ holds for all possible certificates.
(b) If the answer to the input x is "Yes" then $t(x, y) = 0$ holds for at least one certificate.

The Question "P = NP?"   The class P is clearly contained in the class NP. A very important open problem of complexity theory is the question

$$\text{``P} \stackrel{?}{=} \text{NP''}.$$

It belongs to the Millenium Prize Problems listed by the Clay Mathematics Institute, which has offered 1 million US dollars for a solution. The importance of the problem can be explained by the fact that there exist numerous problems for which no polynomial time algorithm is known, but it can be shown that they are contained in the class NP. To determine if there does not exist a polynomial time algorithm for these problems or if such an algorithm exists and is simply not yet found, it is necessary to answer the question "P $\stackrel{?}{=}$ NP".

A decision problem A is called NP-*hard* if every problem in NP can be reduced to A in polynomial time. We call A NP-*complete* if it is additionally contained in NP. For an exact definition of these terms we refer to Garey and Johnson [46].

NP-complete problems are the "hardest" problems in the class NP. We have: If any NP-complete problem can be solved in polynomial time, then so can every other problem in NP and we have P = NP.

An example of an NP-complete decision problem is the question of whether a given finite graph contains a *Hamilton cycle*:

Example C.7 Let G be an (undirected) finite graph. Does there exist a closed path in G that passes through each vertex exactly once?

Almost all experts in the field of complexity theory believe that the classes P and NP are distinct.

The Complexity Class #P   Enumeration problems can be studied analogously to decision problems. The output here is a natural number. In the same way as for optimization problems, there exists a direct relation to decision problems.

Definition C.8 An enumeration problem A is contained in #P if there exists a decision problem $B \in$ NP such that the task of A is to compute the number of solutions that validate B.

Similarly to the terms "NP-hard" and "NP-complete", it is possible to define corresponding classes for enumeration problems. An enumeration problem is #P-*hard* if every problem in #P can be reduced to it. It is called #P-*complete* if it is #P-hard and contained in #P.

Example C.9 The problem of determining the number of different Hamilton cycles in a given finite graph is #P-complete.

Further Complexity Classes   The number of complexity classes which are studied in the literature seems to be continuously increasing. This collection of all such classes is sometimes described as the "zoo" of complexity classes.

In the remarks to Chapter 9 we mention EXPSPACE, the class of algorithms that need at most $\exp^{O(1)}$ storage space.

Appendix D
Software

There exists a large volume of software devoted to the topic of *computational geometry*. The variety ranges from the implementation of single algorithms to large systems with a broad spectrum of applications. This section lists five software packages and their applications regarding computational geometry.

D.1 `polymake`

The system `polymake` specializes in algorithms to study the geometry and combinatorics of polytopes and polyhedra in arbitrary dimension [47, 48]. There are several convex hull algorithms available and Voronoi diagrams and Delone subdivisions can also be computed. In addition to the study of polytopes, the current version 2.12 supplies methods to study matroids, algebraic invariants of finite simplicial complexes as well as algorithms for tropical geometry.

    `polymake` is an open-source system which is written in Perl and C++. Both languages can be used to extend the software. It also offers a substantial C++ library for linear algebra and computational geometry, which can be used independently of the system. The interface is based on a shell which uses a dialect of Perl as its language. Alternatively, `polymake` can be used as a callable library.

    On the Web you can find `polymake` at www.polymake.org.

D.2 `Maple`

`Maple` is a commercial mathematical software system with extensive functionality. The current version 15 provides only a few of the computational geometric algorithms which are discussed in the first part of this book. Specifically, it contains a convex hull algorithm in the plane and a library for solving linear programs. However, `Maple` can compute Gröbner bases and can handle the elimination techniques

M. Joswig, T. Theobald, *Polyhedral and Algebraic Methods in Computational Geometry*,   237
Universitext, DOI 10.1007/978-1-4471-4817-3,

from the second part of the book. Furthermore, Maple provides simple visualization techniques.

Maple has numerous extensions and application examples. A good source for information on these is www.maplesoft.com. Maple defines its own programming language and has interfaces to C and Java.

In comparison to the special foci of the other programs listed here, Maple is often inferior with regard to the scope and speed of its methods. However, it offers the possibility to combine methods from these special areas in one package.

When trying the code examples given in this book, please remember that the syntax of different versions of Maple may vary.

D.3 Singular

Singular is an open-source software project that is dedicated to computational commutative algebra and algebraic geometry [33, 52]. The current version is number 3.1.3. It implements several methods for the computation of Gröbner bases. Elimination and many refinements such as the Conti–Traverso method from Section 10.6 are available. In addition, the system offers algorithms for invariant theory and coding theory as well as numerical methods for solving systems of polynomial equations. Singular has its own programming language.

The webpage is www.singular.uni-kl.de. Singular is included in the mathematical software system Sage, and thus any distribution of Sage also contains Singular.

D.4 CGAL

The "Computational Geometry Algorithms Library" (CGAL) is a broad open-source software system specifically designed for lower dimensional computational geometry [18]. Voronoi diagrams and Delone triangulations are available in many versions and refinements, including the Voronoi diagrams of line segments given in Section 13.1. It also contains a convex hull algorithm for arbitrary dimensions.

The spectrum of applications ranges from arrangements of lines and curves, lattice generation, geometrical data processing and search structures to motion planning [41].

CGAL is a C++ library which is available in its current version 4.0. There are many examples on the webpage www.cgal.org.

D.5 Sage

Sage is a free open-source mathematics software system which combines the power of many existing open-source packages into a common Python-based interface. For the topics in this book, the most important feature is the interface to `Singular`; in particular, all the example computations with Gröbner bases and similar objects could also be done in `Sage`.

The most recent version 5.0.1 can be downloaded from www.sagemath.org. An interface from `Sage` to `polymake` is being developed; a current snapshot is available at https://bitbucket.org/burcin/pypolymake/src.

Appendix E
Notation

The elements of a vector space are usually denoted as column vectors. Although we strictly adhered to this in the first part of the book, we relax this rule in the second and third part to simplify notation.

The table below lists the most important symbols, usually together with a page number that corresponds to its first appearance.

$|M|$	Number of elements in the set M	
$\mathbb{N} = \{0, 1, 2, \dots\}$	Natural numbers	
$\mathbb{Z}$	Integers	
$\mathbb{Q}$	Rational numbers	
$\mathbb{R}$	Real numbers	
$\mathbb{C}$	Complex numbers	
Id	Identity matrix (of suitable dimension)	
$\mathrm{Sym}(M)$	Set of permutations of the set M, symmetric group acting on M	
$\mathrm{sgn}(\sigma)$	Sign of the permutation $\sigma \in \mathrm{Sym}(M)$	
$\mathrm{int}\, M$	Interior of a set $M \subseteq \mathbb{R}^n$	15
$\overline{M}$	Closure of M	15
∂M	Boundary of M	15
$\mathrm{relint}\, C$	Relative interior of a convex set $C \subseteq \mathbb{R}^n$	15
$(K^n)^*$	Dual space of the vector space K^n	
$\mathbb{P}^n_K$	n-dimensional projective space over K	9
$G_{k,n}\, K$	k-th Grassmannian of K^n	198
$\mathrm{lin}\, M$	Linear hull of a subset M of a vector space	
$\mathrm{aff}\, M$	Affine hull	14
$\mathrm{conv}\, M$	Convex hull	14
$[x, y] = \mathrm{conv}\{x, y\}$	Segment between two points $x, y \in \mathbb{R}^n$	
$\mathrm{pos}\, M$	Positive hull	33

M. Joswig, T. Theobald, *Polyhedral and Algebraic Methods in Computational Geometry*, Universitext, DOI 10.1007/978-1-4471-4817-3,
© Springer-Verlag London 2013

$(x_0 : x_1 : \cdots : x_n)^T$	Homogeneous coordinates of a point in projective space	10
$[a_0 : a_1 : \cdots : a_n]$	(Oriented) homogeneous coordinates of a hyperplane	11, 14
$\langle \cdot, \cdot \rangle$	Inner product, Euclidean scalar product	11, 14
$\| \cdot \|$	Euclidean norm	28
$\mathrm{vol}\, M$	n-dimensional volume of $M \subseteq \mathbb{R}^n$	
M°	Polar set of M	28
$\mathcal{F}(P)$	Face lattice of a polytope P	34
$I(V, \mathcal{H})$	Incidence matrix of the double description $(V, \mathcal{H})$	70
$[\mathcal{C}]$	Polyhedral complex, generated by a family $\mathcal{C}$ of polyhedra (with intersection condition)	83
$\mathrm{VR}_S(p)$	Voronoi region of the point p with respect to $S \subseteq \mathbb{R}^n$	81
$\mathrm{VD}(S)$	Voronoi diagram of $S \subseteq \mathbb{R}^n$	84
$\mathcal{P}(S)$	Polyhedron that emerges from $\mathrm{VD}(S)$ as vertical projection	86
$\mathcal{P}^*(S)$	Delone polytope	101
$\mathrm{DS}(S)$	Delone subdivision	103
$\gcd(f, g)$	Greatest common divisor of f and g	
$\mathrm{lcm}(f, g)$	Least common multiple of f and g	
$\deg_x f$	Degree of the polynomial f in the unknown x	
$\mathrm{tdeg}\, f$	Total degree of f	127
$\mathrm{Res}_x(f, g)$	Resultant of f and g with respect to the unknown x	123
$\langle f_1, \ldots, f_t \rangle$	Ideal generated by the polynomials $f_1, \ldots, f_t$	137
$\mathrm{V}(I)$	Affine or projective algebraic variety defined by the ideal I	137
I_k	k-th elimination ideal of I	137, 158
$\mathrm{rem}_\prec(f; g_1, \ldots, g_t)$	Remainder of the multivariate division	139, 143
$\prec_{\mathrm{lex}}$	Lexicographic monomial order	142
$\prec_{\mathrm{glex}}$	Graded lexicographic monomial order	173
$\prec_{\mathrm{grevlex}}$	Graded reverse lexicographic monomial order	144
M_C	Medial axis of the curve C	182
$\lambda_C(p)$	Local feature size of the curve C in point p	184
$\bigwedge^k V$	k-th exterior power of the vector space V	195
$\bigwedge V$	Exterior algebra of the vector space V	195
$x \wedge y$	Exterior product of x and y	195
P, NP, #P	Complexity classes	234

References

1. Adams, W.W., Loustaunau, P.: An Introduction to Gröbner Bases. Graduate Studies in Mathematics, vol. 3. American Mathematical Society, Providence (1994)
2. Aigner, M., Ziegler, G.M.: Proofs from THE BOOK, 4th edn. Springer, Berlin (2010)
3. Althaus, E., Mehlhorn, K.: Traveling Salesman-based curve reconstruction in polynomial time. SIAM J. Comput. **31**(1), 27–66 (2001)
4. Amenta, N., Bern, M., Eppstein, D.: The crust and the β-skeleton: combinatorial curve reconstruction. Graph. Models Image Process. **60**, 125–136 (1998)
5. Arrondo, E.: Another elementary proof of the Nullstellensatz. Am. Math. Mon. **113**(2), 169–171 (2006)
6. Avis, D.: lrslib 4.2. http://cgm.cs.mcgill.ca/~avis/C/lrs.html
7. Avis, D., Bremner, D., Seidel, R.: How good are convex hull algorithms? Comput. Geom. **7**(5–6), 265–301 (1997)
8. Avis, D., Fukuda, K.: A pivoting algorithm for convex hulls and vertex enumeration of arrangements and polyhedra. Discrete Comput. Geom. **8**(3), 295–313 (1992)
9. Awange, J.L., Grafarend, E.W.: Solving Algebraic Computational Problems in Geodesy and Geoinformatics. Springer, Berlin (2005)
10. Basu, S., Pollack, R., Roy, M.-F.: Algorithms in Real Algebraic Geometry, 2nd edn. Algorithms and Computation in Mathematics, vol. 10. Springer, Berlin (2006)
11. Becker, T., Weispfenning, V.: Gröbner Bases. Graduate Texts in Mathematics, vol. 141. Springer, New York (1993)
12. Bertsimas, D., Weismantel, R.: Optimization over Integers. Dynamic Ideas, Belmont (2005)
13. Beutelspacher, A., Rosenbaum, U.: Projective Geometry: From Foundations to Applications, p. 258. Cambridge University Press, Cambridge (1998)
14. Blum, H.: A transformation for extracting new descriptors of shape. In: Whaten-Dunn, W. (ed.) Proc. Symposium on Models for the Perception of Speech and Visual Form, pp. 362–380. MIT Press, Cambridge (1967)
15. Boissonnat, J.-D., Yvinec, M.: Algorithmic Geometry. Cambridge University Press, Cambridge (1998)
16. Brøndsted, A.: An Introduction to Convex Polytopes. Graduate Texts in Mathematics, vol. 90. Springer, New York (1983)
17. Buchberger, B.: Ein Algorithmus zum Auffinden der Basiselemente des Restklassenrings nach einem nulldimensionalen Polynomideal. PhD thesis, Universität Innsbruck (1965)
18. CGAL, Computational Geometry Algorithms Library. www.cgal.org
19. Chan, T.M.: Optimal output-sensitive convex hull algorithms in two and three dimensions. Discrete Comput. Geom. **16**(4), 361–368 (1996)

M. Joswig, T. Theobald, *Polyhedral and Algebraic Methods in Computational Geometry*,
Universitext, DOI 10.1007/978-1-4471-4817-3,
© Springer-Verlag London 2013

20. Chan, T.M., Snoeyink, J., Yap, C.-K.: Primal dividing and dual pruning: output-sensitive construction of four-dimensional polytopes and three-dimensional Voronoi diagrams. Discrete Comput. Geom. **18**(4), 433–454 (1997)

21. Chazelle, B.: An optimal convex hull algorithm in any fixed dimension. Discrete Comput. Geom. **10**(4), 377–409 (1993)

22. Chvátal, V.: Linear Programming. W. H. Freeman and Company, New York (1983)

23. Clarkson, K.L., Shor, P.W.: Algorithms for diametral pairs and convex hulls that are optimal, randomized, and incremental. In: Proc. Fourth Annual Symposium on Computational Geometry, Urbana, IL, 1988, pp. 12–17. ACM, New York (1988)

24. CoCoA-Team: CoCoA: a system for doing Computations in Commutative Algebra. cocoa.dima.unige.it

25. Collins, G.E.: Quantifier elimination for real closed fields by cylindrical algebraic decomposition. In: Automata Theory and Formal Languages, Second GI Conf., Kaiserslautern, 1975. Lecture Notes in Comput. Sci., vol. 33, pp. 134–183. Springer, Berlin (1975)

26. Conti, P., Traverso, C.: Buchberger algorithm and integer programming. In: Applied Algebra, Algebraic Algorithms and Error-Correcting Codes, New Orleans, LA, 1991. Lecture Notes in Comput. Sci., vol. 539, pp. 130–139. Springer, Berlin (1991)

27. Cormen, T.H., Leiserson, C.E., Rivest, R.L., Stein, C.: Introduction to Algorithms, 3rd edn. MIT Press, Cambridge (2009)

28. Cox, D., Little, J., O'Shea, D.: Ideals, Varieties, and Algorithms, 3rd edn. Undergraduate Texts in Mathematics. Springer, New York (2007)

29. Cox, D.A., Little, J., O'Shea, D.: Using Algebraic Geometry, 2nd edn. Graduate Texts in Mathematics, vol. 185. Springer, New York (2005)

30. Crossley, M.D.: Essential Topology. Springer Undergraduate Mathematics Series. Springer, London (2005)

31. de Berg, M., van Kreveld, M., Overmars, M., Schwarzkopf, O.: Computational Geometry, 2nd edn. Springer, Berlin (2000)

32. De Loera, J.A., Rambau, J., Santos, F.: Triangulations. Algorithms and Computation in Mathematics, vol. 25. Springer, Berlin (2010)

33. Decker, W., Greuel, G.-M., Pfister, G., Schönemann, H.: Singular 3.1.3. A computer algebra system for polynomial computations, Universität Kaiserslautern (2011). www.singular.uni-kl.de

34. Dey, T.K., Kumar, P.: A simple provable algorithm for curve reconstruction. In: Proc. Symposium on Discrete Algorithms, Baltimore, MD, pp. 893–894 (1999)

35. Dickson, L.E.: Finiteness of the odd perfect and primitive abundant numbers with n distinct prime factors. Am. J. Math. **35**, 413–422 (1913)

36. Dietmaier, P.: The Stewart–Gough platform of general geometry can have 40 real postures. In: Lenarcic, J., Husty, M.L. (eds.) Advances in Robot Kinematics: Analysis and Control, pp. 7–16. Kluwer Academic, Dordrecht (1998)

37. Dyer, M.E., Frieze, A.M.: On the complexity of computing the volume of a polyhedron. SIAM J. Comput. **17**(5), 967–974 (1988)

38. Edelsbrunner, H.: Algorithms in Combinatorial Geometry. EATCS Monographs on Theoretical Computer Science, vol. 10. Springer, Berlin (1987)

39. Fischer, G.: Plane Algebraic Curves. Student Mathematical Library, vol. 15. American Mathematical Society, Providence (2001)

40. Fischer, G., Piontkowski, J.: Ruled Varieties. Vieweg, Braunschweig (2001)

41. Folgel, E., Halperin, D., Wein, R.: CGAL: Arrangements and Their Applications. Geometry and Computing, vol. 7. Springer, Berlin (2012)

42. Fortune, S.: A sweepline algorithm for Voronoï diagrams. Algorithmica **2**(2), 153–174 (1987)

43. Fukuda, K.: cddlib 0.94b. http://www.ifor.math.ethz.ch/~fukuda/cdd_home/cdd.html

44. Fukuda, K., Prodon, A.: Double description method revisited. In: Combinatorics and Computer Science, Brest, 1995. Lecture Notes in Comput. Sci., vol. 1120, pp. 91–111. Springer, Berlin (1996)

45. Gallier, J.: Discrete Mathematics. Universitext. Springer, New York (2011)

46. Garey, M.R., Johnson, D.S.: Computers and Intractability: A Guide to the Theory of NP-Completeness. Freeman, San Francisco (1979)
47. Gawrilow, E., Joswig, M.: `polymake`: a framework for analyzing convex polytopes. In: Polytopes—Combinatorics and Computation, Oberwolfach, 1997. DMV Sem., vol. 29, pp. 43–73. Birkhäuser, Basel (2000)
48. Gawrilow, E., Joswig, M.: `polymake` 2.12. Technical report, Technische Universität Darmstadt (2012). With contributions by many others, see www.polymake.org
49. Goodman, J.E., O'Rourke, J. (eds.): Handbook of Discrete and Computational Geometry, 2nd edn. Chapman & Hall/CRC, Boca Raton (2004)
50. Gordan, P.: Neuer Beweis des Hilbert'schen Satzes über homogene Functionen. Nachr. Königl. Ges. Wiss. Gött. **3**, 240–242 (1899)
51. Grayson, D.R., Stillman, M.E.: `Macaulay 2`, a software system for research in algebraic geometry. http://www.math.uiuc.edu/Macaulay2/
52. Greuel, G.-M., Pfister, G.: A `Singular` Introduction to Commutative Algebra. Springer, Berlin (2002)
53. Gritzmann, P.: Grundlagen der Mathematischen Optimierung. Springer, Berlin, in preparation
54. Grötschel, M., Lovász, L., Schrijver, A.: Geometric Algorithms and Combinatorial Optimization, 2nd edn. Algorithms and Combinatorics, vol. 2. Springer, Berlin (1993)
55. Gruber, P.: Convex and Discrete Geometry. Grundlehren der Mathematischen Wissenschaften, vol. 336. Springer, Berlin (2007)
56. Grünbaum, B.: Convex Polytopes, 2nd edn. Graduate Texts in Mathematics, vol. 221. Springer, New York (2003)
57. Halperin, D., Kavraki, L., Latombe, J.-C.: Robotics. In: Handbook of Discrete and Computational Geometry, 2nd edn. CRC Press Ser. Discrete Math. Appl., pp. 1065–1094. CRC, Boca Raton (2004)
58. Hatcher, A.: Algebraic Topology. Cambridge University Press, Cambridge (2002)
59. Herstein, I.N.: Topics in Algebra, 2nd edn. Xerox College Publishing, Lexington (1975)
60. Hironaka, H.: Resolution of singularities of an algebraic variety over a field of characteristic zero. I. Ann. Math. (2) **79**, 109–203 (1964)
61. Hironaka, H.: Resolution of singularities of an algebraic variety over a field of characteristic zero. II. Ann. Math. (2) **79**, 205–326 (1964)
62. Hobby, J.: `METAPOST`. http://cm.bell-labs.com/who/hobby/MetaPost.html
63. Hodge, W.V.D., Pedoe, D.: Methods of Algebraic Geometry, vols. i, II. Cambridge University Press, Cambridge (1947)
64. Holzer, S., Labs, O.: `surfex` 0.89. Technical report, Universität Mainz and Universität Saarbrücken (2007). www.surfex.AlgebraicSurface.net
65. Hong, H., Brown, C.W., et al.: `QEPCAD` b 1.46. Technical report, RISC Linz and U.S. Naval Academy, Annapolis (2007). http://www.cs.usna.edu/~qepcad/B/QEPCAD.html
66. Howie, J.M.: Fields and Galois Theory. Springer Undergraduate Mathematics Series. Springer, London (2006)
67. Joswig, M.: Beneath-and-Beyond revisited. In: Algebra, Geometry, and Software Systems, pp. 1–21. Springer, Berlin (2003)
68. Joyce, D.E.: Euclid's Elements. http://aleph0.clarku.edu/~djoyce/java/elements/elements.html (1998)
69. Khachiyan, L., Boros, E., Borys, K., Elbassioni, K., Gurvich, V.: Generating all vertices of a polyhedron is hard. Discrete Comput. Geom. **39**(1–3), 174–190 (2008)
70. Kirwan, F.: Complex Algebraic Curves. London Mathematical Society Student Texts, vol. 23. Cambridge University Press, Cambridge (1992)
71. Klein, R.: Algorithmische Geometrie, 2nd edn. Springer, Berlin (2005)
72. Korte, B., Vygen, J.: Combinatorial Optimization, 3rd edn. Algorithms and Combinatorics, vol. 21. Springer, Berlin (2006)
73. Lang, S.: Calculus of Several Variables, 3rd edn. Undergraduate Texts in Mathematics. Springer, New York (1988)

74. Lang, S.: Undergraduate Algebra, 3rd edn. Undergraduate Texts in Mathematics. Springer, New York (2005)
75. Lazard, D., Merlet, J.-P.: The (true) Stewart platform has 12 configurations. In: Proc. IEEE International Conference on Robotics and Automation, San Diego, CA, pp. 2160–2165 (1994)
76. Mayr, E.W., Meyer, A.R.: The complexity of the word problems for commutative semigroups and polynomial ideals. Adv. Math. **46**(3), 305–329 (1982)
77. McCarthy, J.M.: Geometric Design of Linkages. Interdisciplinary Applied Mathematics, vol. 11. Springer, New York (2000)
78. McMullen, P.: The maximum numbers of faces of a convex polytope. Mathematika **17**, 179–184 (1970)
79. Morris, R.: `SingSurf`: A program for calculating singular algebraic curves and surfaces. www.singsurf.org (2005)
80. Mulmuley, K.: Computational Geometry: An Introduction Through Randomized Algorithms. Prentice Hall, Englewood Cliffs (1993)
81. O'Neill, B.: Elementary Differential Geometry, 2nd edn. Elsevier/Academic Press, Amsterdam (2006)
82. Polthier, K., Preuss, E., Hildebrandt, K., Reitebuch, U.: `JavaView`, Version 3.95. www.javaview.de (2005)
83. Pottmann, H., Wallner, J.: Computational Line Geometry. Springer, Berlin (2001)
84. Preparata, F.P., Hong, S.J.: Convex hulls of finite sets of points in two and three dimensions. Commun. ACM **20**(2), 87–93 (1977)
85. Pressley, A.: Elementary Differential Geometry, 2nd edn. Springer Undergraduate Mathematics Series. Springer, London (2010)
86. Pugh, C.C.: Real Mathematical Analysis. Undergraduate Texts in Mathematics. Springer, New York (2002)
87. Rabinowitsch, J.L.: Zum Hilbertschen Nullstellensatz. Math. Ann. **102**, 520 (1929)
88. Richter-Gebert, J.: Perspectives on Projective Geometry. Springer, Heidelberg (2011)
89. Roman, S.: Advanced Linear Algebra. Graduate Texts in Mathematics, vol. 135. Springer, New York (2008)
90. Santos, F.L.: A counterexample to the Hirsch conjecture. Ann. Math. **176**, 383–412 (2012)
91. Schrijver, A.: Theory of Linear and Integer Programming. Wiley, Chichester (1986)
92. Sottile, F., Theobald, T.: Line problems in nonlinear computational geometry. In: Surveys on Discrete and Computational Geometry. Contemp. Math., vol. 453, pp. 411–432. American Mathematical Society, Providence (2008)
93. Stoer, J., Bulirsch, R.: Introduction to Numerical Analysis, 3rd edn. Texts in Applied Mathematics, vol. 12. Springer, New York (2002)
94. Sturmfels, B.: Gröbner Bases and Convex Polytopes. University Lecture Series, vol. 8. American Mathematical Society, Providence (1996)
95. Sturmfels, B.: Solving Systems of Polynomial Equations. CBMS Regional Conference Series in Mathematics, vol. 97. American Mathematical Society, Providence (2002)
96. Vempala, S.: Geometric random walks: a survey. In: Combinatorial and Computational Geometry. Math. Sci. Res. Inst. Publ., vol. 52, pp. 577–616. Cambridge University Press, Cambridge (2005)
97. von zur Gathen, J., Gerhard, J.: Modern Computer Algebra, 2nd edn. Cambridge University Press, Cambridge (2003)
98. Webster, R.: Convexity. The Clarendon Press/Oxford University Press, New York (1994)
99. Ziegler, G.M.: Lectures on Polytopes. Graduate Texts in Mathematics, vol. 152. Springer, New York (1995)

Index

M. Joswig, T. Theobald, *Polyhedral and Algebraic Methods in Computational Geometry*,   247
Universitext, DOI 10.1007/978-1-4471-4817-3,
© Springer-Verlag London 2013

If you have any concerns about our products,
you can contact us on
ProductSafety@springernature.com
In case Publisher is established outside the EU,
the EU authorized representative is:
**Springer Nature Customer Service Center GmbH
Europaplatz 3, 69115 Heidelberg, Germany**
Printed by Libri Plureos GmbH
in Hamburg, Germany